Ensemble Classification Methods with Applications in R

Ensemble Classification Methods with Applications in R

Edited by
Esteban Alfaro, Matías Gámez and Noelia García
University of Castilla-La Mancha, Spain

WILEY

This edition first published 2019

Registered Offices
John Wiley & Sons, Inc., 111 River Street, Hoboken, NJ 07030, USA
John Wiley & Sons Ltd, The Atrium, Southern Gate, Chichester, West Sussex, PO19 8SQ, UK

Editorial Office
9600 Garsington Road, Oxford, OX4 2DQ, UK

For details of our global editorial offices, customer services, and more information about Wiley products visit us at www.wiley.com.

Wiley also publishes its books in a variety of electronic formats and by print-on-demand. Some content that appears in standard print versions of this book may not be available in other formats.

Library of Congress Cataloging-in-Publication Data

Names: Alfaro, Esteban, 1977- editor. | Gámez, Matías, 1966- editor . | García, Noelia, 1973- editor.
Title: Ensemble classification methods with applications in R / edited by Esteban Alfaro, Matías Gámez, Noelia García.
Description: Hoboken, NJ : John Wiley & Sons, 2019. | Includes bibliographical references and index. |
Identifiers: LCCN 2018022257 (print) | LCCN 2018033307 (ebook) | ISBN 9781119421573 (Adobe PDF) | ISBN 9781119421559 (ePub) | ISBN 9781119421092 (hardcover)
Subjects: LCSH: Machine learning–Statistical methods. | R (Computer program language)
Classification: LCC Q325.5 (ebook) | LCC Q325.5 .E568 2018 (print) | DDC 006.3/1–dc23
LC record available at https://lccn.loc.gov/2018022257

Cover Design: Wiley
Cover Image: Courtesy of Esteban Alfaro via wordle.net

Set in 10/12pt WarnockPro by SPi Global, Chennai, India

Printed in Singapore by C.O.S. Printers Pte Ltd

10 9 8 7 6 5 4 3 2 1

Contents

List of Contributors

Esteban Alfaro, Economics and Business Faculty, Institute for Regional Development, University of Castilla-La Mancha.

Sanyogita Andriyas, Department of Civil & Environmental Engineering, Utah State University, Logan, Utah, USA.

Eva Cernadas, Centro Singular de Investigación en Tecnoloxías da Información (CiTIUS), University of Santiago de Compostela, Santiago de Compostela, Spain.

Mariola Chrzanowska, Faculty of Applied Informatics and Mathematics, Department of Econometrics and Statistics, Warsaw University of Life Sciences (SGGW), Warsaw, Poland.

Davy Cielen, Maastricht School of Management, Maastricht, the Netherlands.

Kristof Coussement, IESEG Center for Marketing Analytics (ICMA), IESEG School of Management, Université Catholique de Lille, Lille, France.

Koen W. De Bock, Audencia Business School, Nantes, France.

Manuel Fernández-Delgado, Centro Singular de Investigación en Tecnoloxías da Información (CiTIUS), University of Santiago de Compostela, Santiago de Compostela, Spain.

Matías Gámez, Institute for Regional Development, University of Castilla-La Mancha.

Noelia García, Economics and Business Faculty, University of Castilla-La Mancha.

Mariusz Kubus, Department of Mathematics and Computer Science Applications, Opole University of Technology, Poland.

Mac McKee, Utah Water Research Laboratory and Department of Civil & Environmental Engineering, Utah State University, Logan, Utah, USA.

María Pérez-Ortiz, Department of Quantitative Methods, University of Loyola Andalucía, Córdoba, Spain.

Wojciech Rejchel, Faculty of Mathematics and Computer Science, Nicolaus Copernicus University, Torun, Poland; Faculty of Mathematics, Informatics and Mechanics, University of Warsaw, Warsaw, Poland.

Dorota Witkowska, Department of Finance and Strategic Management, University of Lodz, Lodz, Poland.

List of Tables

List of Figures

Preface

This book introduces the reader to ensemble classifier methods by describing the most commonly used techniques. The goal is not a complete analysis of all the techniques and their applications, nor an exhaustive tour through all the subjects and aspects that come up within this field of continuous expansion. On the contrary, the aim is to show in an intuitive way how ensemble classification has arisen as an extension of the individual classifiers, and describe their basic characteristics and what kind of problems can emerge from its use. The book is therefore intended for everyone interested in starting in these fields, especially students, teachers, researchers, and people dealing with statistical classification.

To achieve these goals, the work has been structured in two sections containing a total of 11 chapters. The first section, which is more theoretical, contains the four initial chapters, including the introduction. The second section, from the fifth chapter to the end, has a much more practical nature, illustrating with examples of business failure prediction, zoology, and ecology, among others, how the previously studied techniques are applied.

After a brief introduction where the fundamental concepts of statistical classification tasks through decision trees are established, in the second chapter the generalization error is decomposed in three terms (the Bayes risk, the bias, and the variance). Moreover, the instability of classifiers is studied, dealing with the changes suffered by the classifier when it faces small changes in the training set. The three reasons proposed by Dietterich to explain the superiority of ensemble classifiers over single ones are given (statistical, computational, and representation). Finally, the Bayesian perspective is mentioned.

In the third chapter, several classifications of ensemble methods are enumerated, focusing on that which distinguishes between generative and non-generative methods. After that, the bagging method is studied. This uses several bootstrap samples of the original set to train a set of basic classifiers that afterwards combine by majority vote. In addition, the boosting method is analysed, highlighting the most commonly used algorithm, the AdaBoost. This algorithm repeatedly applies the classification system to the training set, but focuses, in each iteration, on the most difficult examples, and later combines the built classifiers through weighted majority vote. To end this chapter, the random forest method is briefly mentioned. This generates a set of trees, introducing in their building process a degree of randomness to assure some diversity in the final ensemble.

The last chapter of the first part shows with simple applications how the individual and ensembled classification trees are applied in practice using the `rpart`, `adabag` and `randomForest` R packages. Moreover, the improvement in the results achieved over individual classification techniques is highlighted.

The second part begins with a chapter dealing with business failure prediction. Specifically, it compares the prediction accuracy of ensemble trees and neural networks for a set of European firms, considering the usual predicting variables such as financial ratios, as well as qualitative variables, such as firm size, activity, and legal structure. It shows that the ensemble trees decrease the generalization error by about 30% with respect to the error produced with a neural network.

The sixth chapter describes the experience of M. Fernández-Delgado, E. Cernadas, and M. Pérez-Ortiz using ensemble methods for classifying texture feature patterns in histological images of fish gonad cells. The results were also good compared to ordinal classifiers with stages of fish oocytes, whose development follows a natural time ordering.

In the seventh chapter W. Rejchel considers the ranking problem that is popular in the machine learning community. The goal is to predict or to guess the ordering between objects on the basis of their observed features. This work focuses on ranking estimators that are obtained by minimization of an empirical risk with a convex loss function, for instance boosting algorithms. Generalization bounds for the excess risk of ranking estimators that decrease faster than a threshold are constructed. In addition, the quality of procedures on simulated data sets is investigated.

In the eighth chapter S. Andriyas and M. McKee implement ensemble classification trees to analyze farmers' irrigation decisions and consequently forecast future decisions. Readily available data on biophysical conditions in fields and the irrigation delivery system during the growing season can be utilized to anticipate irrigation water orders in the absence of any predictive socio-economic information that could be used to provide clues for future irrigation decisions. This can subsequently be useful in making short-term demand forecasts.

The ninth chapter, by M. Kubus, focuses on two properties of a boosted set of rules. He discusses the stability and robustness of irrelevant variables, which can deteriorate the predictive ability of the model. He also compares the generalization errors of SLIPPER and AdaBoost.M1 in computational experiments using benchmark data and artificially generated irrelevant variables from various distributions.

The tenth chapter shows how M. Chrzanowska, E. Alfaro, and D. Witkowska apply individual and ensemble trees for credit scoring, which is a crucial problem for a bank as it has a critical influence on its financial outcome. Therefore, to assess credit risk (or a client's creditworthiness) various statistical tools may be used, including classification methods.

The aim of the last chapter by K. W. De Bock, K. Coussement, and D. Cielen is to provide an introduction to GAMs and GAM-based ensembles, as well as an overview of experiments conducted to evaluate and benchmark the performance, and to provide insights into these novel algorithms using real-life data sets from various application domains.

Thanks are due to all our collaborators and colleagues, especially the Economics and Business Faculty of Albacete, the Public Economy, Statistics and Economic Policy Department, the Quantitative Methods and Socio-economic Development Group

(MECYDES) at the Institute for Regional Development (IDR) and the University of Castilla-La Mancha (UCLM). At Wiley, we would like to thank to Alison Oliver and Jemima Kingsly for their help and two anonymous reviewers for their comments.

Finally, we thank our families for their understanding and help in every moment: Nieves, Emilio, Francisco, María, Esteban, and Pilar; Matías, Clara, David, and Enrique.

1

Introduction

Esteban Alfaro, Matías Gámez, and Noelia García

1.1 Introduction

Classification as a statistical task is present in a wide range of real-life contexts as diverse as, for example, the mechanical procedure to send letters based on the automatic reading of the postal codes, decisions regarding credit applications from individuals or the preliminary diagnosis of a patient's condition to enable immediate treatment while waiting for the final results of tests.

In its most general form, the term classification can cover any context in which a decision is taken or a prediction is made based on the information available at that time, and a classification procedure is, then, a formal method to repeat the arguments that led to that decision for new situations.

This work focuses on a more specific interpretation. The problem is to build a procedure that will be applied to a set of cases in which each new case has to be assigned to one of a set of predefined classes or subpopulations on the basis of observed characteristics or attributes.

The construction of a classification system from a set of data for which actual classes are known has been called different things, such as pattern recognition, discriminant analysis, or supervised learning. The latter name is used rather than unsupervised learning or clustering in which classes are not defined a priori but they are inferred from the data. This work focuses on the first type of classification tasks.

1.2 Definition

The most traditional statistical technique applied to supervised classification is linear discriminant analysis, but in recent decades a wider set of new methods has been developed, in part due to the improvement in the capabilities of informatics supports. Generally, the performance of a classification procedure is analysed based on its accuracy, that is, the percentage of correct classified cases. The existence of a correct classification implies the existence of an expert or supervisor capable of providing it,

Ensemble Classification Methods with Applications in R, First Edition.
Edited by Esteban Alfaro, Matías Gámez, and Noelia García.

so why would we want to replace this exact system by an approximation? Among the reasons for this replacement we could mention:

1. Speed. Automatic procedures are usually quick and they can help to save time. For instance, automatic readers of postal codes are able to read most letters, leaving only some very complex cases to human experts.
2. Objectivity. Important decisions have to be taken basing on objective criteria under the same conditions for all cases. Objectivity is sometimes difficult to ensure in the case of human deciders. In such cases, decisions can be affected from external factors, which would led us to take biased decisions.
3. Explanatory capabilities. Some of the classification methods allow us not only to classify observations but to explain the reasons for the decision in terms of a set of statistical features.
4. Economy. Having an expert who make decisions can be much more expensive than developing an effective classification system from accumulated experience so that it can be applied by anyone, not necessarily expert on the subject but following the guidelines given by the classifier.

1.3 Taxonomy of Supervised Classification Methods

There is no single taxonomy of classification methods, but we can find a variety of them depending on the criterion of division we take, for example between parametric and nonparametric methods or between methods that attempt to estimate probability densities, posterior probabilities or just decision borders. If we consider the first criterion classification methods can be divided into:

- Parametric methods. These methods are based on the assumption of knowing the shape of the underlying density functions, generally the normal distribution. Then the problem is the parameter estimation, which is performed either by maximizing the likelihood or through Bayesian methods. Such methods include Fisher linear discriminant analysis, multiple discriminant analysis, quadratic discriminant, and the expectation–maximization algorithm, among others.
- Non-parametric methods. These methods do not require any hypothesis about the underlying density functions so they are appropriate when the data probability distribution is unknown. These methods include Parzen windows estimation, K-nearest neighbors, classification trees and artificial neural networks.

On the other hand, Lippmann (1991) recognizes five basic types of classifiers:

- Probabilistic methods. These are functional and parametric techniques and therefore indicated when the functional form fits well with the actual distribution of data and there is a sufficient number of examples to estimate parameters. As examples we can point to Gaussian or linear discriminant classifiers based on mixtures of normal distributions.
- Global methods. These are methods that build the discriminant function by internal nodes using sigmoid or polynomial functions that have high non-zero responses over a large part of the input space. These methods include multilayer perceptron, Boltzmann machines, and high-order polynomial networks.
- Local methods. Unlike the previous methods, these techniques build the discriminant function using nodes having nonzero responses only on localized regions of

the input space. Examples of such methods are radial basis functions networks and the Kernel discriminant. The advantage of these methods is that they do not require assumptions about the underlying distributions.

- Nearest neighbour. These methods are based on the distance between a new element and the set of stored elements. Among the best-known techniques are learning vector quantization (LVQ) and K-nearest neighbours. These are non-parametric methods but they require a lot of computing time.
- Rule-based methods. These methods divide the input space into labelled regions through rules or logical thresholds. These techniques include classification trees.

The first three types of methods provide continuous outputs that can estimate either the likelihood or Bayes posterior probabilities, while the last two blocks provide binary outputs. Because of this difference, the first methods respond to a strategy of minimizing a cost function such as the sum of squared errors or entropy, while the second block will aim to minimize the number of misclassified items.

In this work we will focus on the last type of classifiers, to be used as base classifiers in the ensembles. Therefore, the accuracy (error) of the classification system will be measured by the percentage of successes (failures) in the classified elements.

1.4 Estimation of the Accuracy of a Classification System

In the development of a classification system three stages can be set up: selection, training, and validation. In the first stage, both the technique and the set of potential features must be selected. Once the first stage has been completed, it is time to start the learning process through a set of training examples. In order to check the performance of the trained classifier, that is to say its ability to classify new observations in the correct way, the accuracy has to be estimated.

Once the classifier has been validated, the system will be ready to be used. Otherwise, it will be necessary to return to the selection or the training stages, for example modifying the number and type of attributes, the number of rules and/or conjunctions, etc. or even looking for another more appropriate classification method.

To measure the goodness of fit of a classifier the error rate can be used. The true error rate of a classifier is defined as the error percentage when the system is tested on the distribution of cases in the population. This error rate can be empirically approximated by a test set consisting of a large number of new cases collected independently of the examples used to train the classifier. The error rate is defined as the ratio between the number of mistakes and the number of classified cases.

$$error\ rate = \frac{number\ of\ errors}{number\ of\ cases} \tag{1.1}$$

For the sake of simplicity, all errors are assumed to have the same importance, although this might not be true in a real case.

The true error rate could be computed if the number of examples tended to infinity. In a real case, however, the number of available examples is always finite and often relatively small. Therefore, the true error rate has to be estimated from the error rate calculated on a small sample or using statistical sampling techniques (random resampling, bootstrap, etc.). The estimation will usually be biased and the bias has to be analysed in order to find non-random errors. Its variance is important too, seeking the greatest possible stability.

1.4.1 The Apparent Error Rate

The apparent error rate of a classifier is the error rate calculated from the examples of the training set. If the training set is unlimited, the rate of apparent error will coincide with the true error rate, but as already noted, this does not happen in the real world and, in general, samples of limited size will have to be used to build and evaluate a classifier system.

Overall, the rate of apparent error will be biased downwards so the apparent error rate will underestimate the true error rate (Efron, 1986). This usually happens when the classifier has been over-adjusted to the particular characteristics of the sample instead of discovering the underlying structure in the population. This problem results in classifiers with a very low rate of apparent error, even zero in the training set, but with bad generalization ability, that is, bad performance when facing new observations.

1.4.2 Estimation of the True Error Rate

Since the apparent error rate is usually misleading, it is necessary to use other estimators of the true error rate. To perform this estimation and according to the number of examples in the sample, there are two alternative ways: the use of a test set or resampling techniques. The first method is the simplest way to estimate the true error, but it can only be used when the size of the data set is big enough. This method randomly divides the data set into two subsets, one is used to train or build the classifier and the other to test its accuracy.

However, for small or moderately sized samples the use of resampling techniques is recommended (random subsampling, cross-validation, leaving-one-out, or bootstrapping). What these methods do is to repeat many times the random division between training and test sets. All these methods are used to estimate the accuracy of the classifier, but the final classifier is calculated using all the available data set.

1.4.3 Error Rate Estimation Methods

Estimation from a Test Set

This method, referred to as "hold-out", involves dividing the original data set into two groups: the training set and the test set. The classifier is then built using only the training set, leaving out the examples of the test set. Once the classifier has been built, the test set is used to evaluate its performance facing new observations. The error rate of the classifier in the examples of the test set is called the test error rate.

The assignment of each observation to one of the two sets must be done randomly. The proportions are typically of $2/3$ (training) and $1/3$ (test), although in large sets, once the number of test examples exceeds 1000 a higher percentage can be used as training examples.

For large samples, the use of a test set achieves good results. However, for moderate-size sets, this method makes one of the two sets or both have an insufficient number of observations. To solve this problem and avoid the pessimistic bias estimation from the test set, some resampling techniques can be applied, repeating in an iterative way the partition of the data set into training and test sets.

Random Subsampling

This method involves repeating the aforementioned partition in such a way that a new classifier is designed from the training data set generated from each partition. The estimated error rate is computed as the average of the errors of classifiers calculated on the test sets generated independently and randomly. Thanks to this averaging process, random subsampling solves the problem of having to rely on a single partition that may not be representative.

Cross-Validation

This method can be actually considered as a special case of the random subsampling, wherein the original data set is randomly divided into v mutually exclusive subsets of approximately the same size. This method is usually known as v-fold cross-validation, with reference to v folders or data subsets. Every test set is used for assessing the performance of each one of the v classifiers constructed, taking the $v - 1$ remaining subsets as a training set. Therefore, at the end we have v classifiers with their respective test error rates. The error rate estimated by cross-validation will be the average of the v error rates, weighted by the sizes of the subsets if they were of different sizes.

This estimator is optimistically biased with respect to the true error rate. According to the empirical results shown in Kohavi et al. (1995), if v is lower than five, the estimation will be too biased, when v is close to 10, the bias would be acceptable, and, finally, if v is greater than 20, the estimation will almost be unbiased. The author also found that the stratified cross-validation generally has lower bias. Stratification consists of building v partitions such that the proportion of each class in the total set is maintained, especially in terms of the hierarchy of classes since it could happen that the majority class in the original set was relegated to the second place in one of the subsets, which would damage the estimation of the true error rate.

The major advantage of cross-validation with respect to other methods of random subsampling is that every available example is used both in the training process and in the accuracy evaluation. The most usual value for v is 10. This size ensures that the reduction of the training set is not too important in comparison with the proportion for training and test set $(2/3, 1/3)$ usually set in random subsampling. The main drawback is the excessive computational cost due to the v iterations of the training process. When the size of the data set is not big enough, a particular version of cross-validation is recommended, leaving-one-out.

Leaving-One-Out

This method is a particular case of cross-validation in which the whole data set is randomly partitioned into N subsets ($v = N$) in such a way that every subset is formed by only one observation. Then for each one of the N subsets a classifier is generated, trained using the $N - 1$ examples and, finally, assessed with the retained observation.

This way, all the examples are used once as test examples and the remaining cases can be used as training examples in each iteration. The main drawback is that this method is even more expensive in terms of computational cost than 10-fold cross-validation because N classifiers have to be built. For this reason, the leaving-one-out method is particularly recommended when the sample size is small. The error rate is calculated as the numbers of incorrectly classified observations along the N test sets divided into N and this is an asymptotically unbiased estimator of the true error rate.

Bootstrapping

Although the leaving-one-out estimator is asymptotically unbiased, its variance is high for small samples (< 30 cases). To solve this problem, the bootstrapping method can be applied.

As always, the starting point is a set with N examples and the training set, containing N elements, is generated using sampling with replacement, meaning that a single example can be repeated two or more times in a training set. The test set contains those examples that are not included in the training set. The expected proportion of examples included in the training set is 63.2% and the expected proportion of examples in the test set is 36.8%. The estimated error rate is the average of the error rates after several iterations. Around 200 iterations are needed for this estimate to be considered good, therefore this method is computationally more expensive than leaving-one-out.

Weiss and Kulikowski (1991) recommend choosing the resampling method depending on the number of examples available:

- For sample sizes greater than 100, cross-validation is recommended. Although stratified or leaving-one-out cross-validation could also be used, the computational cost would be higher and cross-validation with $v = 10$ yields good results.
- For samples with fewer than 100 examples the leaving-one-out method is advisable.
- For very small samples, fewer than 30 examples, bootstrapping is recommended.

Table 1.1 shows a comparison of the five methods discussed and the recommended choice depending on the number of examples.

Where N is the total number of available examples, n_1 and n_2 are the number of elements in the training and test sets, respectively. This number can vary from 1 to N. B is the number of iterations, which should be much lower than N.

1.4.4 The Standard Error

Although in the previous sections we have focused on the bias, the variability has also to be taken into account to find the estimator accuracy. To do this, standard error (SE) is used. From the binomial distribution the standard error for a set of independent test examples is calculated as

$$SE = \sqrt{\frac{\varepsilon(1-\varepsilon)}{n_2}} \tag{1.2}$$

where ε is the error and n_2 is the number of independent examples used as the test set.

Table 1.1 Comparison of real error rate estimation methods.

	Estimation from a test set	Random subsampling	Cross-validation (10 fold)	Leaving-one-out	Bootstrapping
Training set size	n_1	n_1	$90\%N$	$N-1$	N(j different)
Test set size	$n_2 = N - n_1$	$n_2 = N - n_1$	$10\%N$	1	$N-j$
Number of iterations	1	$B \ll N$	10	N	200
Recommended for	$n_2 > 1000$	N big	$N > 100$	$N < 100$	$N < 30$

If any of the resampling techniques is used, n_2 is N, the total number of examples. However, in this case the estimated standard error will be optimistically biased because the test examples are not completely independent since every example in the sample is used to train the classifier and also to test it.

This is solved when the sample contains about 200 examples or more, achieving a very close estimate of the true standard error.

1.5 Classification Trees

Classification trees are one of the most popular inductive and non-parametric learning methods for supervised classification. The main advantage of this method is its simplicity. In fact, classification trees make a recursive partition of the multidimensional input space formed by the set of characteristics that describe the problem, the sample space.

A classification tree is graphically represented by nodes and branches. Each node symbolizes a point or decision on one of the input features. The initial node is usually referred as the root node. Two or more branches can leave from each node, depending on if the decision in this node is binary or not. Finally, there are terminal or leaf nodes in which the decision on the class assignment is given.

When a new example is presented to the tree, it will be carried along the tree, being filtered by the questions in the nodes. Each test has mutually exclusive and exhaustive outputs, which means that every example will always be assigned to one and only one output. That is, no example will be assigned to two outputs of the same node, but neither will there be any example without output assigned.

Decision trees can be used with both continuous and categorical variables, with two or more categories. If dichotomous variables are used, a binary tree is required. Weiss and Kulikowski (1991, p. 117) use a simple example of a binary classification tree (Figure 1.1), where the classes are C_1 "rising stock market" and C_2 "falling stock market". Nodes 1 (root node) and 2 (the only other non-terminal node) are the test of increasing interest

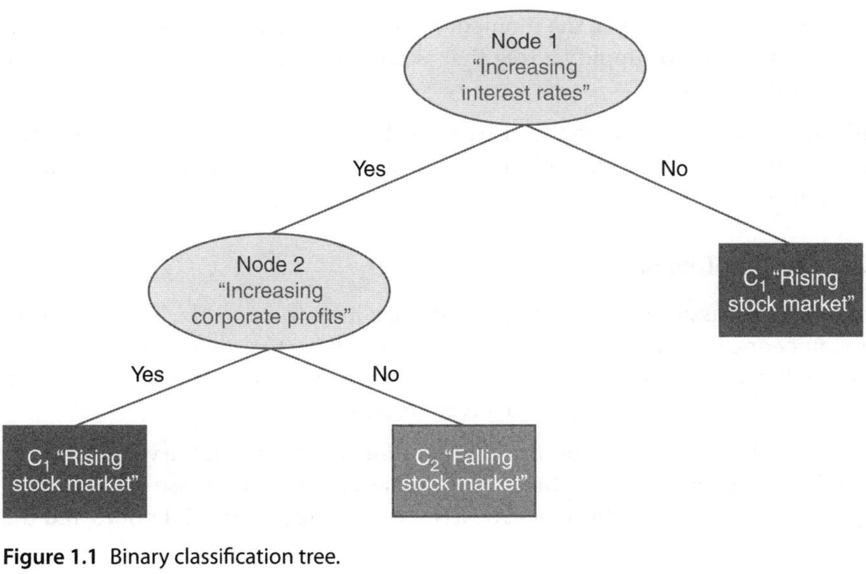

Figure 1.1 Binary classification tree.

rates and the test of increasing corporate profits, respectively. Furthermore, the three terminal nodes or leaves represent three mutually exclusive groups of observations. Two rules cover C_1: non-increasing interest rates or increasing interest rates and increasing corporate profits. One rule covers C_2: increasing interest rates and non-increasing corporate profits.

More formally, we can say that a classification tree, T, represents a recursive partition of the sample space, P, based on a set of examples, S. In addition, the following relationships between the nodes of T, the regions in P, and sets S can be set:

1. Each node of T has associated a subset of examples of S. In the case of the initial node or root, the full set S is assigned.
2. Each leaf, t, has an associated region R_t in P. Thus, if $\tilde{T}$ is the set of leaves of the tree T, then

$$\forall t \in T, \cup R_t = P \tag{1.3}$$

which means that the subsets assigned to the leaf nodes constitute a partition of P. That is, no element in P will remain without an assigned leaf node.
3. Each internal node has an associated region in P, which is the union of the regions associated with the leaf nodes of the subtrees whose root is the same node.
4. The union of the sets of examples assigned to all the nodes on the same level results in the whole set.

There are several ways in which classification trees can be distinguished:

1. The questions in the nodes may be univariate or multivariate, depending on the number of features implied in the decision rule.
2. Internal nodes can have two or more outputs. If all nodes have two outputs, we have a binary tree.
3. The attributes or variables that characterize the examples can be both continuous or categorical.
4. There may be two or more classes. In a two-class problem with dichotomous features, the tree implements a Boolean function and the tree is referred to as a Boolean classification tree.
5. The stopping criterion, that is, the moment from which the tree stops growing.
6. The pruning method to simplify a tree that is too big. The most popular one is the cost-complexity criterion.

The differences among the methods of tree construction will be mainly determined by the latter two issues, the stopping criterion and the pruning method. Both will be studied in later sections.

1.5.1 Classification Tree Building

In order to build a classification tree many decisions have to be made. The analysis of every possibility one by one is a daunting task but we will try to establish a general procedure to be followed. The construction is carried out during the learning phase, which can be outlined in the following steps, which are recursively repeated:

1. Each node is split according to some test, usually formulated about the value of one or more of the characteristics describing the examples. In the binary case (true or false) examples that meet the condition will be assigned to one of the child nodes and the

remaining examples to the other one. Apart from the root node, when a node is split it becomes an internal or intermediate node.

2. The stopping criterion ends the partitioning process. When a node satisfies this condition it is said to be a terminal node or leaf. Examples belonging to a leaf node will have some homogeneity, so that a label with the majority class is assigned to this terminal node. If all examples in a leaf node belong to the same class, it is said to be a pure node.

This is the general scheme followed by all procedures based on trees. In practice many aspects have to be set, such as how splits are made and which is the best split, when to stop the partition process of the nodes and declare them as leaf nodes or which label should be assigned to a leaf node.

1.5.2 Splitting Rule

The splitting rule divides the set of examples into disjoint subsets. The aim is to get subsets which are more homogeneous, in terms of the real class, than the initial set. That is, the subsets will have more purity than the initial set. In order to quantify the homogeneity, an impurity measure will be set. This measure will help to choose the best split and it also will allow the tree to stop growing when a prefixed threshold is reached.

When formulating the questions that will lead to the partitions, the easiest and most commonly used method is to raise binary-type questions. This is the case, for example, for the classification and regression trees (CART) algorithm proposed by Breiman et al. (1984). In CART, Q is the set of all possible binary questions, such as:

$$x \in A?; \quad A \subset P \tag{1.4}$$

where P is the representation space.

Q generates a set of S splits in each node t, and each node t will have two children nodes t_L (left child) and t_R (right child). According to convention:

- Examples in t that fulfil $x \in A$ are assigned to t_L; $t_L = t \cap A$.
- Examples in t that do not fulfil $x \in A$ are assigned to t_R; $t_R = t \cap \overline{A}$.

Each question involves a single attribute. If the attribute is categorical and dichotomous, no transformation is needed since the question will lead to a dichotomous partition. The problem arises when the attribute is categorical with more than two categories or continuous.

In the first case, when the attribute has more than two categories, the solution lies in separating the set of possible categories into two subsets, so that the resulting binary partition remains. For example, if the X_i attribute can take the values {high, low, medium}, possible questions are is X_i high, is X_i low, or is X_i medium?

In case of a continuous-valued attribute, a value that divides its range into two subranges must be set in such a way that examples with a value for X under the threshold is assigned to the left node and the remaining to the right one. In theory any value in the range is valid but it is usual to use the midpoint between two consecutive observed values in the training examples. For example, if X_i is a continuous variable and the values 5, 10 and 15 are observed, the rules $X_i \leq 7.5$ or $X_i \leq 12.5$ could be applied.

1.5.3 Splitting Criteria

Among all the possible splits that could be made, the one selected should lead to the maximum increase in the homogeneity or purity level for descendant nodes with respect to the parent node. Then an impurity/purity measure is needed. According to Breiman et al. (1984) this measure should be zero if the node is pure and take the highest value if all classes have the same proportion of cases.

Among the most popular measures are entropy and a measure based on the Gini index.

The entropy for a node t can be computed as

$$i_E(t) = -\sum_{j=1}^{q} p(j|t)\ln(p(j|t)) \tag{1.5}$$

where it is assumed that $0 \cdot \ln(0) = 0$ and $p(j|t)$ is estimated as the proportion of examples of class j in the node t, that is:

$$p(j|t) = \frac{N_j(t)}{N(t)} \tag{1.6}$$

being $N(t)$ and $N_j(t)$, the total number of examples in node t and the number of examples of class j in t, respectively.

The Gini index measures the concentration or homogeneity of classes in a node and it is computed as

$$i_G(t) = -\sum_{i,\, j=1}^{q} p(i|t) \cdot p(j|t) = 1 - \sum_{j=1}^{q} p(j|t)^2 \tag{1.7}$$

There are several more criteria in the literature, but the final classifier is not usually very sensitive to this election (Breiman et al., 1984).

For the specific case of a two-class problem, the proportions are p and $1-p$, and it is very easy to see that the maximum dispersion is reached when $p = 0.5$.

1.5.4 Goodness of a Split

Whatever the measure of impurity used to evaluate the goodness of a split, the impurity of the node to be split is compared with the impurity of the child nodes. Thus, for a partition (S) of the attribute A, goodness is calculated as the information gain, or decrease in impurity, which is achieved by separating the examples according to the attribute. Specifically, the information gain of an attribute A regarding a set of examples S, $\Delta_i(S, A)$ is calculated as:

$$\Delta_i(S, A) = i(S) - \sum_{v=1}^{k} \frac{N_v}{N} i(S_v) \tag{1.8}$$

where k is the number of possible values of an attribute A and N_v/N is the proportion of examples of S in which attribute A takes the value v and therefore belongs to the subset S_v. The second part in equation 1.8 is the expected value of the impurity after splitting S using attribute A. It is computed through the addition of the impurity for each subset S_v, weighted by the proportion of examples of S that belong to S_v. Hence, the information gain is the expected reduction in impurity after knowing the value of the attribute, that is, the information that attribute A provides on the desired classification.

For the particular case of binary splits, the information gain at node t will be:

$$\Delta_i(t) = i(t) - p_L i(t_L) - p_R i(t_R) \tag{1.9}$$

Among all the possible partitions, the one that maximizes the gain of information and the reduction of the impurity should be chosen. That is, first, for each attribute A_i, the partition is chosen to maximize the information gain, and then, once the optimal partition for each attribute has been obtained, the attribute which provides the highest information gain will be selected.

1.5.5 The Impurity of a Tree

In order to compare different trees that could be built on the same data set, it is interesting to know the average impurity of a given T tree, whose set of leaves can be called $\tilde{T}$, that is calculated as:

$$I(T) = \sum_{t \in \tilde{T}} i(t)p(t) \tag{1.10}$$

where $p(t)$ is the probability of an example reaching the leaf node t. That is, the impurity of a tree is determined by its leaves or terminal nodes.

Breiman et al. (1984) conclude that selecting the partitions that individually maximize the information gain is equivalent to selecting the partitions that minimize the global impurity $I(T)$. That is, the optimal tree will be reached by choosing the best partition for each node. Each time a node t is split, the overall mean impurity reduction is equal to $p(t)$ by the gain of information obtained with the partition, so when the impurity reduction is maximized at t, the average impurity of the tree is also minimized.

Obviously, if there exist two trees T_1 and T_2 of the same size, constructed from the same sample set and the impurities of both $I(T_1)$ and $I(T_2)$ are known, T_1 is preferred if $I(T_1) < I(T_2)$.

1.5.6 Stopping Criteria

The process of growing a tree or dividing or splitting the nodes must stop at some point, but there are several possibilities and the choice of a stopping criterion to optimize the tree obtained is not simple:

1. Continue dividing the nodes until all leaf nodes are pure, that is, until we obtain subsets in which all the examples are of the same class. This can lead to leaf nodes with subsets that are too small, so a second criterion could be needed.
2. Stop splitting nodes when they are pure, or when their size is under a certain threshold. The threshold value will depend on the initial size of the training set, since, for instance, starting from a set of 1000 examples, a node with fewer than 10 cases can be considered small, whereas if the initial set is 50 cases, a node with an equal number of examples will not be too small. The size of the minority class should also be taken into account.
3. Stop the growth of the tree when the leaf nodes exceed a certain level of purity, considered in terms of the proportion of the majority class. For example, the majority class may be required to exceed 70% to declare the node as a leaf node and stop the split. This percentage can be set higher or lower, depending on how demanding you want to be and to balance the average impurity and the size of the tree.

4. Information gain or impurity reduction can also be used as a criterion to stop the tree growth. In this case a threshold $\beta \geq 0$ is also set, so that node t will be a leaf node if

$$\max_{S} \Delta i(S, t) \leq \beta \tag{1.11}$$

   That is, if none of the possible partitions, S, of the node t gets a reduction of the impurity higher than the threshold β, it will be said that the node t is a leaf node. In this case the threshold β is a critical factor because if it is too low, it will allow the tree to grow too much because it will be easy to find a partition that obtains an information gain greater than β. On the other hand, if β is too high, it will be difficult to find a partition whose impurity reduction exceeds the threshold and, therefore, the tree construction will be stopped prematurely. In addition, it is possible that after partitions with small information gains, other partitions could be obtained that achieve a greater reduction in impurity.
5. One last reason to stop the undesirable growth of the tree is to have used all the available information. When there are few characteristics that describe each observation, it could happen that, after exhausting all the information, no leaf nodes have yet been found that meet any of the criteria and it will be mandatory to stop the growth of the tree.

The election of more or less restricted criteria could lead to under- or overfitted trees. Sometimes, the strategy used is to build a very large tree, without stopping its growth, and then proceed to pruning from the leaves to the root (in the opposite direction of its construction), obtaining a decreasing set of trees.

Once a node is declared as a leaf, it is necessary to decide which class will be assigned to the examples that reach that node. If all the training examples of that node belong to the same class C, then obviously that leaf is labeled with class C. In the case that there are examples of several classes, the majority class will be assigned within that leaf t, that is:

$$C(t) = C \text{ if } p(C|t) = \max_{j=1,2,\ldots,q} p(j|t) \tag{1.12}$$

where, as shown in equation 1.6, the probabilities $p(j|t)$ are estimated by the proportion of examples of class j at node t. In the case of a tie between two or more classes, one of them is randomly chosen.

1.5.7 Overfitting in Classification Trees

When a tree is grown until it reaches pure or very small leaves, in the final splits the selection of the most appropriate partition is made on the basis of a very small set of examples, so that there is a risk of overfitting or overlearning. This may be because the training set contains random errors (noise) or regularities that do not come from a true but a spurious relationship between the partition and the classification to be determined.

The existing methods to avoid overfitting when building a classification tree fall into one of two large groups:

1. The methods that stop the growth of the tree before the leaf nodes are pure or too small.
2. The methods that develop the tree until the leaf nodes are pure or very small and then apply some pruning mechanism to the whole tree.

A priori, the first method may seem faster and better, but Breiman et al. (1984) defend a posteriori pruning instead of the interruption of growth. According to them pruning is more efficient, since it allows one or several subtrees of one node to remain and the rest disappear while, if growth is interrupted, all branches are eliminated.

Weiss and Kulikowski (1991) propose using cross-validation techniques to estimate the real error of the tree over the whole population, since the error in rewriting is an optimistically biased estimator and can continue to decrease until reaching zero when all the leaves are pure. However, the real error stops decreasing at a certain point, remains a constant size for a number of slightly higher leaf nodes, and then begins to grow as the tree increases (see Figure 1.2). What is proposed is to stop the growth of the tree when the real error reaches its minimum. That is, if a partition does not decrease the real error, then that partition is not made and the node is declared terminal. The resulting tree is a subtree of the entire tree and its error will not be less than half the real error of the whole tree. That is, at best you can improve the real tree error by 50%.

To determine the correct size of the tree, several criteria can be used:

1. Use a different training set to evaluate the effectiveness of pruning the tree nodes.
2. Use all the examples available for training and apply a statistical test to decide if pruning a node produces an improvement for the training set. Quinlan (1986) uses a χ^2 test to estimate whether the division of a node improves behavior over the sample total population or only over the sample of the training set.
3. Use an explicit measure of complexity to encode training examples and the classification tree, stopping the growth of the tree when minimizing the size of the encoding. These methods are based on the principle of minimum length of description (Quinlan and Rivest, 1989).

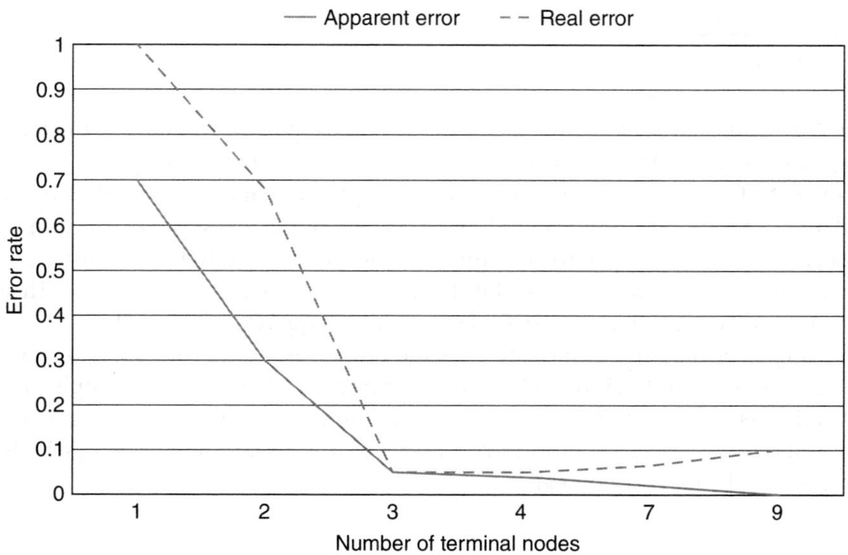

Figure 1.2 Evolution of error rates depending on the size of the tree.

1.5.8 Pruning Rules

As discussed, it is more efficient to develop the tree completely and prune it later. The general procedure is as follows:

1. Develop the tree until all the leaves are pure or have a very small size. This will result in an oversized tree being represented by T (complete tree).
2. Once the complete tree T is obtained, it is pruned to achieve a decreasing and nested sequence of trees. If T_1 is the result of pruning T, T_1 is said to be a pruned subtree of T and is represented as $T_1 \prec T$, so that the complete sequence of pruned and nested trees can be represented as $T \succ T_1 \succ T_2 \succ T_3 \succ \cdots \succ t_1$, where t_1 is a tree with a single node.

It is therefore a question of choosing one of these trees. To do this an error measure will be assigned to each of the trees, finally selecting the tree with the lowest error. It is assumed that smaller trees will have smaller errors when evaluating them on new observations, since they will not be over-adjusted and will have a greater generalization capability, against more complex trees that will tend to be over-adjusted to the examples of the training set and will obtain worse results for new observations.

Cost-Complexity Pruning

The method of cost-complexity pruning is also known as weakest link pruning. It prunes backwards from the complete tree. To avoid having to consider all possible pruned subtrees, only the subtrees obtained by pruning the weakest branch in the tree are considered. This branch is determined by considering at the same time the error committed and the size of the tree to be represented by $|\tilde{T}|$. Specifically, the weakest branch will be that subtree rooted in a non-terminal node t, which can be erased with the smallest increase in error per pruned node. At each node t we quantify the effect of pruning the rooted subtree at t by $g(t)$, which is calculated as:

$$g(t) = \frac{R(t) - R(T_t)}{|\tilde{T}|} \tag{1.13}$$

where $R(t)$ is the error at node t.

The node t, the elimination of which is less damaging to the performance of the tree (minimum $g(t)$), is pruned and the process is repeated for rest of the tree.

This pruning strategy includes a parameter that has preference for the simplest trees versus the larger ones, that is, it penalizes the complexity of the trees. In fact the aim is a balance between the accuracy of the classification of the tree and the size of the tree, so that a small increase in error is allowed if this leads to a large reduction in size. This penalty factor is a threshold that if increased results in a sequence of nested trees.

The cost-complexity measure of the tree T[1], which is represented by $R_\alpha(T)$, is an adequately weighted linear combination of the cost or error of the T tree and its complexity. The classification error of T is represented by $R(T)$, and the complexity of T means its size or number of leaves. The complexity parameter α is a real value ($\alpha \geq 0$), which represents the complexity cost per leaf.

---

1 Although the estimation of the classification error as the cost of the tree is used in this paper, the entropy or the deviation of the partition can also be used.

Specifically, the measure of cost-complexity is calculated as

$$R_\alpha(t) = R(t) + \alpha|\tilde{T}| \tag{1.14}$$

A large value of α will mean a high cost per leaf and will favor subtrees with fewer terminal nodes. Therefore, for each value of α one has to find the tree $T(\alpha)$ nested with the whole tree T that minimizes $R_\alpha(T_i)$, that is, to obtain

$$R_\alpha(T(\alpha)) = \min_{T_i < T}\{R_\alpha(T_i)\} \tag{1.15}$$

Thus, if we progressively increase α from $\alpha = 0$, we will generate a finite and unique sequence of nested subtrees $T_1 > T_2 > T_3 > \cdots > t_1$, where $T_k = T(\alpha_k)$ and $\alpha_1 = 0$.

The specific procedure for obtaining the above sequence of trees can be found in Breiman et al. (1984). In general terms, a tree with many leaves is represented by T_1 ($\alpha_1 = 0$), the weakest branch of T_1 is searched for and pruned, creating T_2 when α reaches α_2. Then the weakest branch of T_2 is searched for and pruned, creating T_3 when α reaches α_3, and so on until the tree constituted only by the root node is reached. As α grows, trees become smaller and smaller, resulting in a decreasing sequence of nested subtrees.

The optimal tree must be chosen from the previous ones, and will be the one that has the lowest value for the measure of cost-complexity, that is, T_{opt} will be chosen such that

$$\hat{R}(T_{opt}) = \min_{\forall k} R(T_k) \tag{1.16}$$

For considerations on the estimation of classification error, see section 1.4.

The 1-SE Rule

Breiman et al. (1984), instead of choosing the minimum $R(T_k)$, proposed the 1-SE rule, according to which one must choose the smallest tree whose error does not exceed the minimum error plus one standard deviation.[2] Specifically the 1-SE rule will is as follows: if T_{k_0} is the tree such $R(T_{k_0}) = \min_k R(T_k)$, then the selected tree will be T_{k_1}, where k_1 is the maximum k that satisfies:[3]

$$R(T_{k_1}) \le R(T_{k_0}) + SE(T_{k_0}) \tag{1.17}$$

On the one hand, the 1-SE rule reduces the instability associated with the selection of the exact minimum while on the other hand selecting the simplest tree whose goodness is comparable to the minimum $R(T_k)$. This rule is easily understood by looking at Figure 1.3, which relates the cross-validation error, the cost-complexity measure of a tree, and its number of leaves. According to Breiman et al. (1984), the evolution of tree error is characterized, in general, by a rapid initial descent, which stops in a flat zone (valley), more or less prolonged to finally ascend when the size of the tree is high.

Pruning of Rules

This method of pruning is used by the well-known C4.5 algorithm, proposed in Quinlan (1993)[4] and consists of converting the fully developed tree into a set of rules, one for each leaf, and later simplifying this set of rules. The steps to follow are:

---

2 The standard deviation of the error is known as the standard error. This is why it is represented by SE and the rule is represented by 1-SE.

3 The maximum k is chosen because it will be the one that gives the simplest tree.

4 The system C4.5 was born from the development of the original ID3 system of the same author.

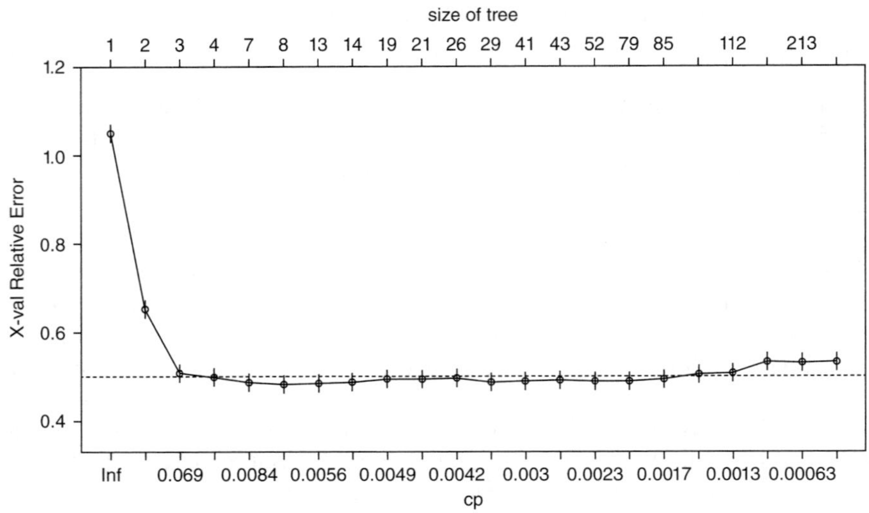

Figure 1.3 Evolution of cross-validation error based on the tree size and the cost-complexity measure.

1. The tree that best fits the training set is built, without worrying about whether or not there is overfitting.
2. The learned tree is converted into a set of rules equivalent to the tree, creating a rule for each path from root to leaves, that is, there will be as many rules as leaf nodes.
3. Prune each rule, eliminating any antecedents of the rule as long as the estimated accuracy is improved.
4. A contribution to this method of pruning is to introduce a fourth step before the last one, in which the pruned rules are ordered by their estimated precision and the precision of the set of rules is checked, leaving aside one of them from lowest to greatest precision. If the precision does not decrease, the rule is removed and the process is followed with the rest. If the precision worsens, the rule is added back to the set and testing continued with the rest.
5. The rules that have not been eliminated will form the final set of rules. When applying them, the order is preserved from greater to lesser precision.

In this method, a rule is generated for each leaf of the entire tree. Each variable evaluated along the path from the root to the leaf becomes an antecedent of the rule, and the class that labels the leaf node becomes the consequent of the rule.

Then, for each rule, any antecedent whose elimination does not worsen the estimated accuracy of the rule is removed. For rules with two or more antecedents, the accuracy is estimated after removing each one of the antecedents and the one that produces the greatest increase in the precision is eliminated. Subsequently, the process is repeated to decide if a second attribute is deleted, and so on until there is no antecedent whose deletion does not worsen accuracy. No antecedent will be pruned if this improves the estimated accuracy of the rule.

In order to estimate the precision of the rule, in addition to the methods discussed in section 1.4, Quinlan (1993) proposed the use of a pessimistic modification of the apparent or in rewriting error to compensate for the optimistic bias of this estimate. For this purpose, the upper limit of the 95% confidence interval for error estimation is collected, calculating its standard deviation assuming a binomial distribution. Therefore, the value used for the estimation will be the error in rewriting plus 1.96 times the estimated standard deviation.[5] This solution greatly facilitates the calculations, decreasing the high computational cost of the resampling methods seen in section 1.4.

Pruning the rules instead of pruning the tree directly has three main advantages:

1. Since the same inner node belongs to several rules, working with rules allows the different contexts in which a decision node is used to be distinguished and treated independently in each rule. On contrary, if the tree is pruned, the decision would be to prune the node or not.
2. The distinction between the evaluated attributes near the root and those evaluated near the leaves is eliminated.
3. Legibility is improved and rules are often more easily understood by humans.

5 The value 1.96 corresponds to the value of the normal distribution $N(0, 1)$ for the 95% confidence level.

2

Limitation of the Individual Classifiers

Esteban Alfaro, Matías Gámez, and Noelia García

2.1 Introduction

In this chapter, some aspects related to the behavior and properties of individual classifiers will be analyzed. The focus will be on some of the problems or difficulties derived from the use of these classifiers, such as lack of accuracy or instability.

The chapter is structured into four sections, in addition to the introduction. First, in section 2.2 we study the generalization error of individual classifiers. This error will be disaggregated into the sum of three non-negative terms that are Bayes' risk, bias, and variance. The first term collects the error inherent in the data set that no classification method can reduce. The bias measures the persistent error of the classification method, that is, the error that would be maintained even if there were an infinite set of classifiers, independently trained. The term corresponding to the variance measures the error caused by the fluctuations that occur when generating an individual classifier. Therefore, from this approach the idea is that by averaging several classifiers one can reduce the variance and, thus, also decrease the expected or generalization error.

Second, in section 2.3 the issue of instability of some classifiers is addressed. A classifier is unstable if it undergoes major changes for small modifications in the training set. In this sense classification trees and neural networks, for example, are considered as unstable methods, whereas the linear discriminant analysis and the nearest-neighbor method, among others, are considered stable. However, when the size of the training set is too small compared to the data dimension, all classifiers suffer from instability problems. That is, under these circumstances even the linear discriminant is unstable.

In general, the more stable classifiers behave better, whereas the more unstable classifiers obtain worse prediction results. Therefore, the study of stability is interesting in order to improve the accuracy of a classifier. In this sense, a measure of instability is proposed, which reflects how the changes in the composition of the training set affect the classifier under construction.

Section 2.4 analyzes some reasons that explain the superiority of the ensemble methods in comparison to the individual classifiers. For example, assuming that the error rate of each classifier is less than 50% and that the individual classifiers are independent, the error committed by the combination of those classifiers can be reduced to zero when the number of basic classifiers approaches infinity. On the other hand, the ideal combination

Ensemble Classification Methods with Applications in R, First Edition.
Edited by Esteban Alfaro, Matías Gámez, and Noelia García.

uses very accurate classifiers but they should disagree as many times as possible since, in other case, the combination of identical classifiers does not bring any benefit.

Following this reasoning, three reasons are proposed to justify the use of combined classifiers, one statistical, one computational and the third, of representation. The first reason is based on the elimination of the risk associated with the choice of one classifier instead of the others, especially in small size sets where the estimated accuracy may be similar for several classifiers. The computational reason raises the possibility that the classification systems that perform a local search can be trapped in a local optimum and be unable to reach the global optimum. Finally, the representation reason argues the impossibility of representing the true function with any of the available hypotheses.

To conclude this chapter, in section 2.5, a brief commentary on the Bayesian perspective on model aggregation is included, which also advocates this solution when the objective is not the explicit selection of a model, but to extract the best possible performance from the data set available for the prediction to be performed.

2.2 Error Decomposition: Bias and Variance

To study the generalization error of a classifier and to analyze its decomposition in several terms, including bias and variance, we start as in any classification problem with a training set **T**, which contains n examples of k known classes. For each example, the values of p attributes or characteristics, which may be numeric or categorical, are known and form the vector $\mathbf{x} = (x_1, x_2, x_3, \dots, x_p)$ and its y class. From **T**, a classifier $C(\mathbf{x}, \mathbf{T})$ is constructed, from which bias, variance, and prediction error are going to be studied. For simplicity, among the different options available as loss functions, the one chosen here is that limited to count the errors committed by the classifier $C(\mathbf{x}, \mathbf{T})$. The sum of squares of the errors is among the alternative loss functions (Tibshirani, 1996b).

Assuming that the observations $(\mathbf{x_i}, y_i)$ in the training set **T** are a random sample of the true F distribution, usually unknown in practice,

$$(\mathbf{x_i}, y_i) \overset{i.i.d.}{\sim} F, \quad i = 1, \dots, n, \tag{2.1}$$

the prediction error of the $C(\mathbf{x}, \mathbf{T})$ classifier is defined as

$$PE(Y, C) \equiv E_{\mathbf{T}} E_F(C(\mathbf{x}, \mathbf{T})) \neq y) \tag{2.2}$$

where $E_{\mathbf{T}}$ refers to the expectation on the training set **T**, whose elements are independent and identically distributed (i.i.d.) from F, and E_F represents the expected value on the test set $\tilde{\mathbf{T}}$, which is also generated from F. Efron and Tibshirani (1995) differentiate this expected error rate from the conditional error rate that sets the training set **T**. Although the conditional error rate may often be interesting, Efron and Tibshirani (1995) show that its estimation is very difficult, so that in general the unconditional error rate is used.

Once the prediction error concept has been defined, the decomposition of this error is sought in several terms in order to facilitate its study. Before dealing with the decomposition of the error in classification problems, a similar analysis is collected for the case of regression problems, where it was firstly developed.

For regression problems, in which the sum of the squares of the errors is usually used as a loss function, the prediction error can be easily decomposed. Thus, if Y is a continuous random variable

$$Y = C_0(\mathbf{x}) + \varepsilon \tag{2.3}$$

where $\mathrm{E}(\varepsilon|\mathbf{x}) = 0$. Given $\sigma^2(\mathbf{x}) = Var(Y|\mathbf{x})$, then, for an estimator $C(\mathbf{x}, \mathbf{T})$ with $C_A(\mathbf{x}) = \mathrm{E}_\mathbf{T}[C(\mathbf{x}, \mathbf{T})]$, it can be seen that

$$\begin{aligned}
&PE(Y, C(\mathbf{x}, \mathbf{T})) = E_\mathbf{T}E_F[Y - C(\mathbf{x}, \mathbf{T})]^2 = \\
&= E_\mathbf{T}E_F[(Y - C_A(\mathbf{x})) - (C(\mathbf{x}, \mathbf{T}) - C_A(\mathbf{x}))]^2 = \\
&= E_\mathbf{T}E_F[(Y - C_A(\mathbf{x}))^2 + (C(\mathbf{x}, \mathbf{T}) - C_A(\mathbf{x}))^2 - 2(Y - C_A(\mathbf{x}))(C(\mathbf{x}, \mathbf{T}) - C_A(\mathbf{x}))] = \\
&= E_\mathbf{T}E_F[Y - C_A(\mathbf{x})]^2 + E_\mathbf{T}E_F[C(\mathbf{x}, \mathbf{T}) - C_A(\mathbf{x})]^2
\end{aligned} \tag{2.4}$$

on the other hand

$$\begin{aligned}
&E_\mathbf{T}E_F[Y - C_A(\mathbf{x})]^2 = E_F[Y - C_A(\mathbf{x})]^2 = E_F[(C_0(\mathbf{x}) - C_A(\mathbf{x})) + \varepsilon]^2 = \\
&= E_F[(C_0(\mathbf{x}) - C_A(\mathbf{x}))^2 + \varepsilon^2 + 2(C_0(\mathbf{x}) - C_A(\mathbf{x}))\varepsilon] = \\
&= E_F[C_0(\mathbf{x}) - C_A(\mathbf{x})]^2 + E_F[\varepsilon^2]
\end{aligned} \tag{2.5}$$

and using the definitions

$$\begin{aligned}
&Bias^2(C) = E_F[C_A(\mathbf{x}) - C_0(\mathbf{x})]^2 = E_F[Y - C_A(\mathbf{x})]^2 - E_F[\varepsilon^2] \\
&Var(C) = E_\mathbf{T}E_F[C(\mathbf{x}, \mathbf{T}) - C_A(\mathbf{x})]^2
\end{aligned} \tag{2.6}$$

The decomposition is obtained as

$$PE(Y, C(\mathbf{x}, \mathbf{T})) = E_F[\varepsilon^2] + Bias^2(C) + Var(C) \tag{2.7}$$

Therefore, in regression problems the prediction error can be decomposed into the sum of three non-negative terms. The first assumes a lower limit on the prediction error of any regression method and is interpreted as that part of the explained variable variability that cannot be explained by any regression method. It would be something like the intrinsic error to a data set. The second component is known as bias and it shows the extent to which the regression procedure has approximated the objective function, which would achieve this minimum error. Finally, the third term measures how much the predictions of the regression procedure used vary with respect to the average.

A similar decomposition can be obtained in classification for the error measures that are generally used. First, the ideal classifier is defined by $C^*(\mathbf{x}) \equiv E_F[Y|\mathbf{x}]$, known as the optimal Bayes classifier, which is obtained as follows

$$C^*(\mathbf{x}) = \arg\max_{y_j} P(y_j|\mathbf{x})$$

That is, the optimal Bayes classifier, given an observation $\mathbf{x}$, assigns that class y_j that has a greater probability of occurring conditioned to the values taken by observation $\mathbf{x}$.

The aggregate, ensemble or combined classifier is defined as $C_A(\mathbf{x}) \equiv E_\mathbf{T}[C(\mathbf{x}, \mathbf{T})]$. To understand what this ensemble classifier means, suppose that there exists an infinite sequence of training sets and the classifier $C(\mathbf{x}, \mathbf{T})$ is applied to each of them. The aggregate classifier $C_A(\mathbf{x})$ would be the mean of $C(\mathbf{x}, \mathbf{T})$ at $\mathbf{x}$ on this infinite group, if the output of $C(\mathbf{x}, \mathbf{T})$ is continuous (a posteriori probabilities), or the most frequently predicted class if the classifier simply assigns a class. The combined classifier reduces the error of $C(\mathbf{x}, \mathbf{T})$ by adding it over the training sets obtained from F.

Note that the aggregate classifier $C_A(\mathbf{x})$ may vary depending on whether $C(\mathbf{x}, \mathbf{T})$ returns a posteriori probabilities or only assigns the winning class. For example, considering a two-class problem, if for an observation $\mathbf{x}$, $C(\mathbf{x}, \mathbf{T})$ assigns the probabilities $(0.45, 0.55)$ and $(0.9, 0.1)$ with probabilities $2/3$ and $1/3$, respectively, then $C_A(\mathbf{x}) = (0.6, 0.4)$ and, therefore, predicts the first class. But if $C(\mathbf{x}, \mathbf{T})$ gives only the class, i.e. $(0, 1)$ or $(1, 0)$ with probabilities $2/3$ and $1/3$, then $C_A(\mathbf{x}) = (1/3, 2/3)$ and will therefore predict the second class.

Breiman (1996a) uses the term aggregated and calls its bootstrap estimation the bagging classifier (**B**ootstrap and **agg**regat**ing**). This classifier is calculated as $\hat{C}_A(\mathbf{x}) = E_{\hat{F}}[C(\mathbf{x}, \mathbf{T})]$. Bagging imitates the aggregated classifier by averaging $C(\mathbf{x}, \mathbf{T})$ on the training sets obtained from the F estimate. This estimation is performed by generating bootstrap replicas of the original training set. In Breiman (1996a) it is shown that bagging reduces the classification error by at least 20% on the group of problems used.[1] In addition, Breiman gets very small differences when bagging works with estimates of the odds of the classes, or with the class labels. The bagging method will be discussed in more detail in the next chapter.

From the definitions of the optimal Bayes classifier, $C^*(\mathbf{x})$, and the aggregated classifier, $C_A(\mathbf{x})$, we can now define the bias and variance for classifier $C(\mathbf{x}, \mathbf{T})$, which is briefly represented by C, as follows:

$$\begin{aligned} Bias(C) &= PE(C^*, C_A) \\ Var(C) &= PE(C, C_A) \end{aligned} \tag{2.8}$$

Based on these definitions, the following properties are derived:

1. Bias is actually a kind of quadratic error and therefore never negative.
2. The classifier C is unbiased if its aggregated version, C_A, predicts the same class as the optimal Bayes classifier, with probability 1 over the input variables $\mathbf{x}$.
3. The variance is never negative.
4. If the classifier C is deterministic, that is, it does not depend on the training set, then $C = C_A$ and its variance is zero.

The second property asserts that $C(\mathbf{x}, \mathbf{T})$ is unbiased at $\mathbf{x}$ if $C_A(\mathbf{x}) = C^*(\mathbf{x})$. That is, $C(\mathbf{x}, \mathbf{T})$ is unbiased on $\mathbf{x}$ if, in the replicates of $\mathbf{T}$, $C(\mathbf{x}, \mathbf{T})$ chooses the correct class more frequently than any other class. A classifier that is unbiased on $\mathbf{x}$ is not necessarily an infallible classifier. For example, suppose that in a dichotomous problem $P(1|\mathbf{x}) = 0.9$; $P(2|\mathbf{x}) = 0.1$ and $P_{\mathbf{T}}(C(\mathbf{x}, \mathbf{T}) = 1) = 0.6$; $P_{\mathbf{T}}(C(\mathbf{x}, \mathbf{T}) = 2) = 0.4$. Then C is unbiased on $\mathbf{x}$ because it chooses the correct class with more probability than the rest, but the probability that C gets the correct classification is $0.6 \times 0.9 + 0.4 \times 0.1 = 0.58$, while the Bayes classifier C^* has a 0.9 probability of predicting the correct class.

The definition of $Var(C) \equiv PE(C, C_A)$ is equivalent to $PE(C_A, C)$, but it is more natural, since it allows the aggregate version of classifier C to be defined as

$$C_A(\mathbf{x}) = \arg\ \min_{C'} PE(C, C') \tag{2.9}$$

1 In his article "Bagging Predictors", Breiman uses seven sets of the University of California Irvine (UCI) database, all of which are real except for one that is simulated (waveform). Each set is randomly divided into a training and a test set. The individual tree (CART) is constructed in the usual way in the training set and to build the bagging classifier 50 bootstrap replicas of this set are made. This process is repeated 100 times for all seven problems. The comparison between CART and bagging is performed using the mean of the errors over the 100 replicates.

Table 2.1 Error decomposition of a classifier (Tibshirani, 1996a).

Bayes risk	The optimal Bayes classifier error
Bias	Measures how much the aggregated classifier is separated from the optimal Bayes classifier
Variance	Measures how much the individual classifier is separated from the aggregated classifier

Considering the decomposition of the prediction error obtained for the regression problems in equation 2.5 and using the definitions of bias and variance of a classifier that have just been pointed out, Tibshirani (1996a) proposes the following decomposition of the prediction error of a classifier:

$$PE(C) = PE(C') + Bias^2(C) + Var(C) \tag{2.10}$$

For a fixed goal function and training set size, the conventional decomposition formulation divides the expected error into the sum of three non-negative quantities (see Table 2.1):

- **Objective noise (intrinsic)**. This amount is the lowest limit on the expected error of any learning algorithm. It is the expected error of the optimal Bayes classifier and is also known as Bayes risk.
- **Bias^2**. This amount measures how close the learning algorithm approaches on average to the target (on all possible training sets of the same size).
- **Variance**. This amount gathers how much the learning algorithm predictions fluctuate for the different training sets of the given size.

Note that building an aggregate classifier and replacing C with C_A reduces the variance to zero, but there is no guarantee that the bias will be reduced. In fact, some examples could be given in which the bias will increase, as proposed by Tibshirani (1996a): Suppose that $y = 1$ for all $\mathbf{x}$, and the classifier C predicts $y = 1$ (for any $\mathbf{x}$) with probability 0.4 and predicts $y = 0$ (for any $\mathbf{x}$) with probability 0.6. Then $PE(C) = 0.6$ and $PE(C_A) = 1$, so in this case, the prediction error of the aggregate classifier will be higher than that of the individual classifier.

2.3 Study of Classifier Instability

After seeing the decomposition between bias and variance of the generalization error, in this section we study the instability of the classifiers. Even in the big data era, it is common to find small data sets in classification problems, especially in the case of real-world problems, what is known as the problem of small sample size. This happens when the size of the training set is small compared to the dimension of the data, i.e. the number of characteristics that describe each example. In these cases, it may be difficult to construct a good classifier so, for example, in the case of the linear discriminant it would be difficult to construct a good discriminant function, since the estimates of class mean and covariance matrix may be biased. Constructing the classifier using these estimates can lead to a discriminant function that behaves poorly. In addition, when the number of training examples is lower than the dimension of the data, some classifiers cannot be constructed, generally those using the inverse of the covariance matrix. Under these conditions, the sample estimate of the covariance matrix will be a singular matrix, which cannot be inverted. In any case, classifiers built in small training sets are usually biased and may have a large variance. As a result, it is said that these classifiers are unstable.

To study the instability of classifiers, it is necessary to establish some way of measuring this concept. Intuitively, the instability of a classifier can be related to the standard deviation of its classification error in the test set, if available, or the training set in the worst scenario. In this way, we can see how much the classification error varies when considering different training sets of the same size, or different initial values of the parameters used in the construction of a classifier, for example the initial weights in a neural network. However, the standard deviation of the classification error does not directly consider the size of the training set or certain changes its composition. In fact, the standard deviation of the classification error does not show the instability of a classifier. It only shows the instability of its behavior (the classification error) in the available data set (training or test set). The classification error does not necessarily reflect the instability of the classifier, for example in the case of the discriminant analysis the classification error does not show the instability of the discriminant function.

When both the size of the training set and the dimension of the data are small, very different classifiers can be obtained in training sets of the same size, although the behavior of these classifiers may be the same. Also, when both the size of the training set and the dimension of the data are large, the conditional classification error may be the same, but the classifiers will be different. Therefore, to measure the instability of a classifier and see how it varies, another procedure should be used that takes into account the size of the training set and its composition. This is especially important when it is necessary to measure the instability of classifiers built, for example, in bootstrap replicas of the training set.

In Skurichina (2001) a possible measure of the instability of a classifier is proposed. This instability measure aims to consider the effects of changes in the composition of the training set on the response of the classifier. To make the changes in the training set, the aforementioned bootstrap method is used. This technique starts from a set of n elements and consists of selecting n elements drawn with replacement; in this way some of the elements in the original set will appear several times in the bootstrap replica while others do not appear any time. Therefore, starting from the training set $\mathbf{T} = \{(\mathbf{x}_1, y_1), ..., (\mathbf{x}_n, y_n)\}$, bootstrap replicas of type $\mathbf{T}_b = \{(\mathbf{x}_{b1}, y_{b1}), ..., (\mathbf{x}_{bn}, y_{bn})\}$ are made. It is important to note that in $\mathbf{T}_b$ some examples will be present one, two, or even more times, and others may not appear at all in that bootstrap replica.

The measure of instability proposed by Skurichina (2001) is calculated as follows. Let $\mathbf{T} = \{(\mathbf{x}_1, y_1), \dots, (\mathbf{x}_n, y_n)\}$ be the training set and $\tilde{\mathbf{T}} = (\tilde{\mathbf{x}}_1, \tilde{\mathbf{x}}_2, \dots, \tilde{\mathbf{x}}_m)$ the test set.

1. Construct the classifier C in the training set $\mathbf{T}$ and classify the test set $\tilde{\mathbf{T}} \rightarrow C(\tilde{\mathbf{x}}_i)$, where $C(\tilde{\mathbf{x}}_i) = y_i \quad (i = 1, \dots m)$, if $\tilde{\mathbf{x}}_i$ is classified by C in the class y_i.
2. Repeat for $b = 1, 2, \dots, B$.
   a) Take the sample bootstrap $\mathbf{T}_b$ and construct the basic classifier C_b in this bootstrap sample and classify the test set $\tilde{\mathbf{T}} \rightarrow C_b(\tilde{\mathbf{x}}_i)$, where $C_b(\tilde{\mathbf{x}}_i) = y_i$ for $(i = 1, \dots m)$, if $\tilde{\mathbf{x}}_i$ is classified by C_b in the class y_i.

b) Calculate the probability that the prediction will be different when the test set is classified using C and C_b, $P(C_b(\tilde{\mathbf{x}}_i) \neq C(\tilde{\mathbf{x}}_i))$.

3. Instability is calculated as

$$instability = \frac{1}{B} \sum_{b=1}^{B} P(C_b(\tilde{\mathbf{x}}_i) \neq C(\tilde{\mathbf{x}}_i)) \quad (2.11)$$

Skurichina (2001) shows the advantages of this measure of instability compared to the standard deviation of the classification error, using artificially generated data sets. Measuring the instability of a classifier through the standard deviation of the mean classification error, whether obtained in the test set, generalization error, or in the same training set, apparent error, does not provide satisfactory results in the case where the training set is small. The standard deviation of the mean generalization error is misleading when the classifiers, obtained in different training sets of small size, perform equally poorly on the test set. However, these classifiers may actually be very different from one another, for example in the case of classification trees, different trees may have a similar behavior. In this case, the standard deviation of the mean generalization error will be relatively small, although the classifier is very unstable.

Likewise, the standard deviation of the mean generalization error can lead to confusion when we have distinct classifiers that behave very well, in the most extreme case all of them with zero apparent error. Although the classifier is unstable, the standard deviation of the apparent error will be equal to zero, which would indicate a high stability for the classifier. The measure of instability in equation 2.11 does not present these drawbacks, being more objective to measure classifier instability.

It can be seen that the behavior of the classifiers and their stability are correlated. As a general rule, the most unstable classifiers have a poorer performance and the most stable classifiers perform better. Therefore, it is logical to think that to improve the behavior of an unstable classifier, it must be stabilized.

According to Breiman (1996a), classification trees and neural networks are unstable methods, while the nearest-neighbor and linear discriminant methods are stable. Methods such as trees have a high variance, but on average they are right, i.e. are quite unbiased, the correct class is usually the winner if the majority vote is applied for the aggregation of several of them. Stable methods, such as the linear discriminant, achieve their stability by having a very limited set of models that fit the data. The result is a low variance, but if the data cannot be correctly represented with the set of available models, a high bias can be obtained.

To calculate the instability measure of a classifier (equation 2.11), Skurichina (2001) uses 25 bootstrap replicates of the training set and constructs a basic classifier in each of them. This process is repeated 50 times in independent training sets of the same size to obtain the average of the results.

Theoretically the measure of instability must use the test set, since in this way the result is more objective than the instability measured in the training set. Instability shows how much the classifier is affected by changes in the composition of the training set. When considering the instability measured in the test set, for any set of data, the classifiers reach maximum instability for small sizes of the training set. If few observations are available for learning, even small changes in the composition of the training set can greatly affect the resulting classifier, for example the classification tree or the discriminant function. If you increase the number of training examples above the critical size given by the dimension of the data, the classifiers become more stable and their generalization error decreases. The composition of the training set is less important when the size of the training set is large enough to correctly represent the true distribution of the problem.

The instability measured in the training set is more biased because the same training set is used to construct the classifier and also to estimate its instability. However, in practice the test set cannot be used and one of the techniques set out in the first chapter for the estimation of the error will have to be used. In addition, it can be observed that the behavior of instability in relation to the sample size is similar whether measured in the training set or in the test set. That is, the instability measured in the training and test sets fluctuates almost in parallel, decreasing as the number of observations increases.

In general, the instability of a classifier will depend on several factors such as the complexity of the classifier, the distribution of the data used to build it, and the sensitivity of the classifier to the size and composition of the training set.

2.4 Advantages of Ensemble Classifiers

In this section, some justifications are given for the superiority of the combined classifiers over individual classifiers. The starting idea is to use a set of classifiers to obtain a greater precision than each one of them achieves individually. Each classification method is based on different estimation concepts or procedures. In addition, since all sorting methods have some strength or advantage over the default rule, it is logical to try to combine the best properties of each of them by combining them in some way. Even combining classifiers based on the same procedure is beneficial, as will be seen in Chapter 3. To combine the outputs of the individual classifiers, we sometimes work with the predicted class, while in others the probabilities assigned to each class by each of them are used.

Combining the output of several classifiers is useful only if there is disagreement between them. Obviously combining several identical classifiers does not get any benefit. Hansen and Salamon (1990) proved that if the mean error rate for one observation is less than 50% and the classifiers used in the ensemble are independent in producing their errors, the expected error for one observation can be reduced to zero when the combined number of classifiers tends to infinity.

Later, Krogh and Vedelsby (1995) proved that the joint error can be divided into a term that measures the mean generalization error of each individual classifier and another that collects disagreement between classifiers. What they formally demonstrated was that the ideal combination consists of classifiers with high precision, which disagree among them as many times as possible.

Going back to the Hansen and Salomon approach, even for the simplest method of combination, the majority vote, and assuming that the classifiers are independent of each other, we can check for a dichotomous problem how the accuracy of the set is superior to that of the individual classifiers, provided that they make an error of less than 50%. Specifically, it can be seen as the probability that more than half of all individual classifiers considered fail. Figure 2.1 shows how the ensemble error, as a function of the number of classifiers used in the combination, evolves for different values of the error

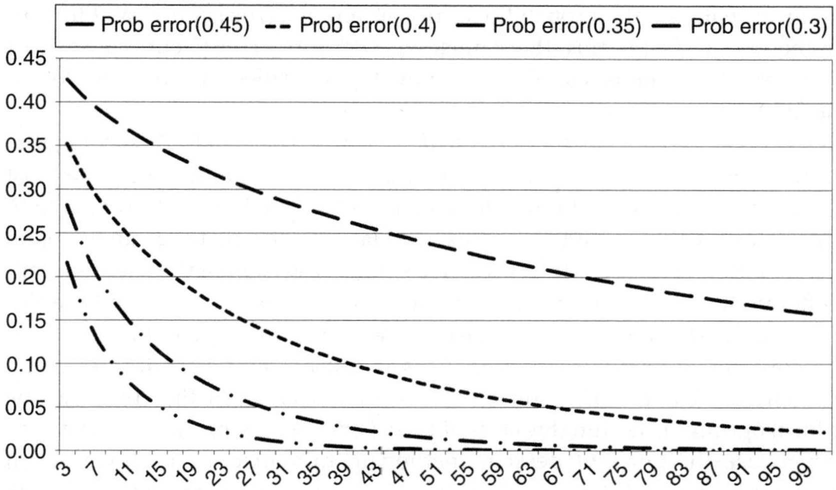

Figure 2.1 Probability of error depending on the number of classifiers in the ensemble.

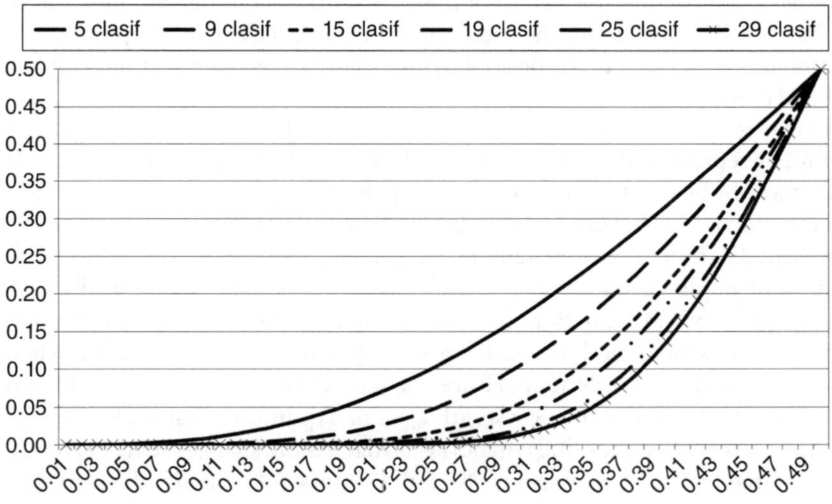

Figure 2.2 Probability of error of the ensemble depending on the individual classifier accuracy.

probability of the individual classifiers (ε). Only combinations with an odd number of classifiers are considered, as is often done in practice to avoid draws. As the number of classifiers used increases, the ensemble accuracy increases for a given value of individual accuracy. Figure 2.2 shows how the ensemble error varies, depending on the accuracy of the base classifiers, between 0 and 0.5, for different sizes of the combination. As might be expected, the lower the accuracy of the basic classifiers, the smaller the ensemble accuracy if its size is fixed. Whatever the number of classifiers used, when its accuracy reaches 0.5, no benefit is obtained from this combination.

Despite the simplicity and clarity of this explanation, it is based on very restrictive assumptions in practice, such as independence between the basic classifiers. In addition, the approach for calculating the probability of the ensemble error is only valid for this combination method. For all of this, there must be other more consistent reasons that explain the superiority of the ensemble over individual classifiers. Dietterich (2000a) considers the following three reasons.

The first reason is referred to as the *statistical reason* by Dietterich. A learning system can be understood as the search, within a certain space of hypotheses, of the most adequate one. In practice, it is common to find data sets that are too small compared to the size of the hypothesis space. In these cases, the classification method can be found with several different hypotheses, which present a similar precision in the data set available. If the different hypotheses are combined, the results can be averaged and the risk of choosing a classifier that obtains less precision when generalizing is reduced.

Second, he considers a *computing reason*. This is caused by the nature of certain classification systems that perform some type of local search, so that they can be trapped in a local optimum. Within this class of classifiers are, among others, the neural networks, which use the gradient descent method to minimize a cost function on the training observations, and the classification trees, which use a sequential splitting rule in the development of the tree. Even in cases where the size of the training set is enough big so that the above problem does not exist, it is still computationally difficult for the learning system to find the best hypothesis. In conclusion, the combination of classifiers, obtained by performing the local search from different initial points, will achieve a better approximation to the desired hypothesis or objective function than any of the individual classifiers.

The third reason is the *representation reason*. In many classification problems, the true function cannot be represented by any of the hypotheses existing in the space of available hypotheses. However, by combinations of relatively simple hypotheses of that space, a better approximation to the real function can be reached since it is possible to extend the space of representable functions, for example using a weighted sum of simple assumptions. This question applies even to those classification methods that theoretically have the capacity to represent any function, for instance neural networks or classification trees, which are very flexible methods with a great capability for representation when the number of observations is unlimited. However, in practice it is usual to face limited data sets, so that even these systems work with a set of hypotheses that is also limited and will stop the search when they find a hypothesis that fits the available data correctly.

These are the three fundamental reasons Dietterich points out to justify the superiority of the ensemble classifiers over the individual. However, it should not be forgotten that they also present some disadvantages, such as loss of comprehensibility (more complex structure and larger size), slower construction of the combination (greater memory capacity is required), and taking more time in classifying a new observation.

2.5 Bayesian Perspective of Ensemble Classifiers

To conclude the chapter, the Bayesian view on the aggregation of classification methods is collected, or more generally on the aggregation of models. From a Bayesian perspective, the aggregation of models is the most natural option. In this way, not only the risk or uncertainty inherent in the data set is taken into account, but also the risk associated with the choice of a single model among the various options available.

First, it is necessary to distinguish whether the purpose is the explicit selection of a model, to see which one presents more suitable characteristics for a given task, or, on the contrary, the objective is the prediction in future observations from what is learned during training in the available data set. Thus, when the purpose is not to select a particular model but to obtain maximum performance from the available data set, the most appropriate option from a Bayesian point of view is the aggregation or combination of models. This in Bayesian terminology is known as Bayesian model averaging, which considers the uncertainty due to the models. For a more detailed description see Hoeting et al. (1999). In particular, this technique acts as follows. If y is the class to be predicted, its posterior distribution given the set of data $\mathbf{T}$ will be:

$$P(y|\mathbf{T}) = \sum_{b=1}^{B} P(y|M_b, \mathbf{T})P(M_b, \mathbf{T}) \tag{2.12}$$

In this way, the a posteriori distributions are averaged under each of the considered models, weighted by the a posteriori probability of each model. In this equation $M_1, M_2, \ldots, M_B$ are the considered models.

This general framework of Bayesian averaging of models can be applied equally to the case of interest in this work: the classification methods. In this case, the models $M_1, M_2, \ldots, M_B$ are the classification procedures, which are combined to reduce the risk associated with the choice of an individual method.

3

Ensemble Classifiers Methods

Esteban Alfaro, Matías Gámez, and Noelia García

3.1 Introduction

In the previous chapter we studied some aspects related to the behavior and properties of individual or basic classifiers, such as bias, variance, and instability. In addition, some reasons have been given that explain the superiority of the combined classifiers over the individual classifiers.

This chapter deals with the study of combined classifiers, focusing mainly on the boosting method, but also collecting a classification of the ensemble methods and introducing the bagging and random forests methods. The structure of the chapter is discussed below.

First, some of the taxonomies of classifier combination methods are collected. Due to the importance of aggregation of methods in pattern recognition, several classifications of methods of combining learning systems have been proposed since the mid 1990s. Some taxonomies of these methods are made based on certain characteristics that they present, such as whether the combined classifiers use the same representation or not for input observations, the aggregation architecture, if they select the basic classifiers or simply combine them, or if the method generates new individual classifiers or just combines the existing ones.

Next, the bagging approach is analyzed, which is based on the aggregation of a set of basic classifiers, where each one is trained on a bootstrap replica of the original training set and the final classification decision is based on the majority vote of these basic classifiers. This system takes advantage of the bootstrap resampling technique, as well as the aggregation, hence its name, bagging, which is an acronym of "**b**ootstrapping and **agg**regat**ing**".

Third, the boosting method, which aims to improve the accuracy of a given classification system, is described. To do this, we first train a classifier whose accuracy in the training set exceeds at least that of the default rule (0.5 in the dichotomous case with equal a priori probabilities). Subsequently, new basic classifiers are added to form a combination whose accuracy in the training set is arbitrarily high. In that case, it is said that the behavior of the classifier has been boosted. In a schematic way, boosting can be considered to train successively the basic classifiers with modified versions of the training set, which are more informative given the existing set of basic classifiers. The final classification of an observation is based on the outputs of the basic classifiers.

Ensemble Classification Methods with Applications in R, First Edition.
Edited by Esteban Alfaro, Matías Gámez, and Noelia García.

Finally, to conclude the chapter, the random forest method is presented, where a group of trees is generated for its later combination. To show that these trees are not related to each other, randomness is introduced in the generation of them so that each tree will be a function of the training set, but also of a random vector, which will influence its development.

3.2 Taxonomy of Ensemble Methods

Due to the importance that has been given to the aggregation of methods in pattern recognition since the mid 1990s, several classifications of methods of combining learning systems have been proposed in this period. Although there are others, only those considered most important have been selected here.

First, we consider the classification proposed by Kittler et al. (1998) that basically distinguishes two combination scenarios according to the point of view of the analysis. In the first type, all classifiers use the same representation in input patterns or observations, for example a set of nearest-neighbor classifiers using the same vector of variables, but differing in the value of k or in the distance used to determine the number of neighbors to consider, or a set of neural networks with the same structure, but with different vector of weights obtained from different learning strategies. In this way, each classifier, for a given observation, can be understood to obtain an estimate of the same a posteriori probability of a class.

In the second case, each classifier uses its own representation of input observations, that is, the measurements taken from the observations are unique for each classifier. An important application of aggregation of classifiers in this scenario is the integration of different types of characteristics or attributes of observations. In this situation, it is no longer possible to consider the a posteriori probabilities calculated as an estimate of the same functional value, since the classification systems work in different measurement spaces.

In Dietterich (2000a), the following groups of combination methods are established, differentiating between those that perform a Bayesian type of voting, those that modify training examples, those that modify variables, those that modify possible classes, and lastly, those that randomize the learning system.

Third, Lam (2000) proposes to group these methods according to the architecture of the aggregation, distinguishing if it is done in series, in parallel or hierarchically. In Jain et al. (2000) the separation is done according to whether the basic classifiers are selected or not by the combination algorithm, differentiating between selection-oriented combination methods and those oriented to the combination.

Finally, Valentini and Masulli (2002) use a classification similar to the previous one, depending on whether or not the combination algorithm acts on the basic classifiers by modifying them. In this way, they distinguish between generative and non-generative methods of combination. Generative methods create sets of basic classifiers by acting on the classification system itself, or on the set of training data, actively seeking to improve the diversity and accuracy of the basic classifiers. Non-generative methods are limited to combining a given set of possibly well-designed basic classifiers, that is, they do not generate new basic classifiers but attempt to combine, in the best possible way, those that already exist. A more detailed reference to this classification is given below.

3.2.1 Non-Generative Methods

In non-generative methods, the basic classifiers are joined by a combination procedure that depends on its adaptability to input observations and the needs of the output provided by individual learning systems, that is, the type of combination depends on the type of output. If only the assigned class is available, or if the continuous outputs are difficult to handle, then the majority vote is used. This is the simplest way to combine several classifiers and consists of assigning an observation to the class that most classifiers predict. That is, each new observation must be presented to each individual classifier to see what class it is assigned to. Once all the individual classifiers have given their prediction, the most commonly repeated class is assigned to that new observation. In the case of a tie, the a priori most likely class is assigned or, if it is unknown, the majority class in the training set or, in the worst case, it is solved at random.

The problem is that in this way all classifiers of the committee have the same importance and the greater or lesser generalization accuracy of them is not considered. In addition, the number of classifiers included will be a critical factor, since too many inaccurate classifiers may lead to a wrong final decision and, therefore, to a decrease in the accuracy of the ensemble.

This procedure can be improved by assigning a weight to each individual classifier so that the behavior of the combined classifier in the training set is optimized. If the basic classifiers provide the a posteriori probabilities of the classes, they can be added using simple operators such as minimum, maximum, mean, median, product or weighted average.

3.2.2 Generative Methods

Generative methods attempt to improve the overall accuracy of the combination by directly acting on the accuracy and diversity of the base classifiers. Within this group, subgroups can be distinguished, depending on the strategies used to improve basic classifiers.

1. Resampling methods. Resampling techniques can be used to extract different samples from the original data set so that the individual classifiers trained in each of them are subsequently used in the combination. Among these techniques, the one most commonly used for this purpose is the bootstrap technique, which allows samples with replacement of the same size as the original data set to be obtained, creating different training sets. These techniques are especially advisable for unstable classification systems, that is, methods that are very sensitive to small changes in the training set, such as classification trees and neural networks. Within these combination methods that use resampling the most commonly used are the bagging and boosting methods, which will be discussed later.
2. Variable selection methods. These reduce the number of features used in basic classifiers. In this way, the dimensionality problem is dealt with (Skurichina and Duin, 2001, p. 156), which consists of a shortage of examples in relation to the number of variables describing each of the observations. These feature subspaces methods act by dividing the attribute set, using each subset to train a basic classifier. The most commonly used method is the random subspace method proposed by Ho (1998), which randomly selects a subset of characteristics where the base classifier is trained. In this

way, a random subspace of the original feature space is obtained and the classifier is constructed in that subspace. Aggregation is usually done by voting weighted by the accuracy of the individual classifiers. In Skurichina and Duin (2001) it is shown that this method is effective for classifiers whose learning curve is decreasing, and which are constructed in small and critical training sets.

3. Testing and selection methods. These are based on the idea of selecting the basic classifiers during the process of creating the combination. Although there are alternatives, it is important to highlight the forward and backward selection, in imitation of the strategies followed for the selection of variables in some techniques, such as discriminant analysis. This consists of a sequential process in which with each step a new classifier of the committee is included (or extracted) only if this entails a reduction in the error.
4. Random aggregation methods. These procedures randomize the learning algorithm to generate combinations of learning systems. An example would be to start with random values for the weights in the backpropagation algorithm in a multilayer perceptron, obtaining different classifiers to use in the combination. Some experimental results show that randomizing the learning systems used to generate the basic combination classifiers improves the behavior of non-random individual classifiers (Dietterich, 2000a). Among these methods are the random forests, which will be discussed in section 3.5.

3.3 Bagging

One of the most common problems in establishing a classifier for a data set is the limited size of the set of training examples. Although this problem particularly affects parametric methods, it is a challenge for any classifier. The smaller the data set available, the less certain it is that this set will faithfully represent the total population. In general, classifiers constructed in small sets may be biased and have a high variance in the probability of misclassification. In this case, the classifier is said to be unstable.

In many cases no further observations are available and therefore the training set is limited. There are several techniques that attempt to obtain more stable classifiers, and this is currently one of the open fields of research in the area of classification systems.

One possible solution is to use the bagging system. In the same way that the bootstrap estimation of the data distribution parameters is more precise and robust than the traditional estimation, one can think of the use of this technique to obtain, once combined, a classifier with better properties.

Bagging was proposed by Breiman (1996a) and is based on bootstrapping and aggregation methods. Both bootstrapping and aggregation methods have beneficial properties. Bootstrapping consists of obtaining random samples with replacement of the same size as the original set.

Starting from the training set $\mathbf{T}_n = \{(\mathbf{x}_1, y_1), \dots, (\mathbf{x}_n, y_n)\}$, where y_i takes values at $1, 2, \dots, k$, B bootstrap samples are obtained ($\mathbf{T}_b$), with $b = 1, 2, \dots, B$. These bootstrap replicates are obtained by drawing with replacement the same number of elements as in the original set (n), In some of these replicates, the presence of noisy observations has been eliminated or at least reduced, as there is a lower proportion of noisy to non-noisy examples. Therefore, the classifiers constructed in these sets will present a better behavior than the classifier built in the original set. Thus, bagging can be really useful for building a better classifier when the training set presents noisy observations.

The combined classifier often obtains better results than the individual classifiers used to construct the final classifier. This can be understood since combining the individual classifiers also combines the advantages of each one in the final classifier.

In particular, the bagging method is applied as follows. A basic classifier $C_b(\mathbf{x}_i)$ is constructed in each $\mathbf{T}_b$ replica, where $C_b(\mathbf{x}_i)$ just as y_i takes values at $1, 2, \ldots, k$. Then the B basic classifiers constructed are combined using the majority vote, so that the most frequent predicted class is the one chosen in the final decision rule:

$$C_f(\mathbf{x}_i) = \underset{j=\{1,2,\ldots,k\}}{\arg\max} \sum_{b=1}^{B} \mathbf{I}(C_b(\mathbf{x}_i) = j) \tag{3.1}$$

The function $\mathbf{I}(i, j)$ is the indicator function, which outputs 1 if the inner expression is true and 0 otherwise. Algorithm 3.1 reflects the bagging method schematically.

Algorithm 3.1 Bagging

1. Repeat for $b = 1, 2, \ldots, B$.
   a Take a bootstrap replicate $\mathbf{T}_b$ of the training set $\mathbf{T}_n$.
   b Construct a single classifier $C_b(\mathbf{x}_i) = \{1, 2, \ldots, k\}$ in $\mathbf{T}_b$.
2. Combine the basic classifiers $C_b(\mathbf{x}_i)$, $b = 1, 2, \ldots, B$ by the majority vote:

$$C_f(\mathbf{x}_i) = \underset{j=\{1,2,\ldots,k\}}{\arg\max} \sum_{b=1}^{B} \mathbf{I}(C_b(\mathbf{x}_i) = j)$$

There are other combination functions based on a posteriori probabilities: minimum, maximum, mean, median, and product. Another possibility is to average the coefficients of the basic classifiers, for which the average function is used. This has the following advantages: first, it is not necessary to retain all the results of the classification, only the coefficients; second, it obtains a single classifier with the same number of parameters as the basic classifiers.

When a bootstrap replica of the initial training set $\mathbf{T}_n = \{(\mathbf{x}_1, y_1), \ldots, (\mathbf{x}_n, y_n)\}$ is drawn, the probability that the ith observation $\mathbf{x}_i$ $(i = 1, 2, \ldots, n)$ is included m times $(m = 0, 1, 2, \ldots, n)$ in that sample bootstrap, $\mathbf{T}_b$, will be given by the binomial distribution $\mathrm{B}(n;1/n)$, where $1/n$ is the probability of $\mathbf{x}_i$ being selected in each extraction and n is the number of extractions with replacement. Therefore, the probability that $\mathbf{x}_i$ is extracted m times will be

$$P(m) = \binom{n}{m} \left(\frac{1}{n}\right)^m \left(1 - \frac{1}{n}\right)^{n-m} \tag{3.2}$$

When $1/n < 0.1$, that is, for sets with more than 10 observations, the binomial distribution can be approximated through the Poisson distribution, $P(\lambda)$, where $\lambda = n(1/n) = 1$, and therefore the probability that $\mathbf{x}_i$ is selected m times is:

$$P(m) = \frac{e^{-1}}{m!} \tag{3.3}$$

Thus, each observation has an approximate probability of $1/e$ of not being included (replacing in the previous equation $m = 0$) in a bootstrap replica. Thus, in the long term it can be expected that, on average, approximately 37% of the observations will be left out of a bootstrap sample. In this way, the possible noisy observations of the training set will not appear in some of these samples. In that case, the classifier built under those conditions will get a better error rate than the built-in original training set with noisy observations. That is why the classifier obtained by bagging obtains better results than the individual classifiers.

In fact, the classifiers built in the bootstrap samples of the training sets sometimes get a better classifier than the original and in some cases a worse classifier. Therefore, the combination of the bootstrap versions of the classifier allows a better classifier than the original to be obtained.

Different versions of bagging have been studied by several researchers, though perhaps more frequently in the area of regression than from the point of view of classification. In Breiman (1996a) it is shown that bagging can reduce the error, as much of regression as of classification, in the decision trees. According to this author, bagging is beneficial only for unstable classifiers, since mainly it manages to reduce the variance, while for stable procedures it could even be detrimental to the classifier's behavior, for example for the nearest-neighbor classification method.

Skurichina and Duin (2001) argue that the stability of linear classifiers depends on the size of the training set. Therefore, it could not be said that bagging is not useful for a particular classifier in general, but will depend on the particular application faced.

3.4 Boosting

As already mentioned, given a set of data, a learning system generates a classifier capable of predicting the class of a new observation. The greater or lesser accuracy of that classifier will depend on the quality of the method used and the difficulty of each task. Whenever the obtained classifier outperforms the default rule, it means that it has been able to find some structure in the data to achieve this advantage. Boosting is a method that increases the accuracy of a classifier by making maximum use of its advantage. Thus, the classification method is used as a subroutine to produce a highly precise classifier in the training set.

Boosting applies the classification system repeatedly on the training data, but in each step the learning attention is focused on different examples of this set using adaptive weights, $w_b(i)$. Once the process has finished, the single classifiers obtained are combined into a final, highly accurate classifier in the training set. The final classifier usually achieves a high degree of accuracy in the test set, as various authors have shown both theoretically and empirically (Banfield et al., 2007; Bauer and Kohavi, 1999; Breiman, 1998; Drucker and Cortes, 1996; Dietterich, 2000a; Friedman et al., 2000; Freund and Schapire, 1996, 1997; Ho, 1998; Quinlan, 1996; Schapire, 1999).

There are several versions of the boosting algorithms (Schapire and Freund, 2012; Friedman et al., 2000), but the most widely used is the one by Freund and Schapire (1996), which is known as Adaboost. For simplification purposes, it can be assumed that there are only two classes without loss of generality. A training set is given by $\mathbf{T}_n = \{(\mathbf{x}_1, y_1), \dots, (\mathbf{x}_n, y_n)\}$, where y_i takes the values $\{-1, 1\}$. The weight $w_b(i)$ is assigned to each observation $\mathbf{x}_i$ and is initially set to $1/n$. This value will be updated after each step. A basic classifier denoted $C_b(\mathbf{x}_i)$ is built on this new training set and is applied to each training example. The error of this classifier is represented by ε_b and is calculated as

$$\varepsilon_b = \sum_{i=1}^{n} w_b(i)\mathbf{I}(C_b(\mathbf{x}_i) \neq y_i) \tag{3.4}$$

From the error of the classifier in the bth iteration, a constant α_b is calculated and this value is used to update the weights. More specifically, according to the authors mentioned above, $\alpha_b = \ln((1 - \varepsilon_b)/\varepsilon_b)$ and the new weight for the $(b+1)$th iteration will be

$$w_{b+1}(i) = w_b(i)\exp(\alpha_b \mathbf{I}(C_b(\mathbf{x}_i) \neq y_i)) \tag{3.5}$$

The calculated weights are then normalized so that they add up to one. According to Freund and Schapire the error should outperform the default rule, $\varepsilon_b = 0.5 - \gamma_b$, where γ_b shows the advantage of the basic classifier of the bth step over the default rule in the worst case scenario, in which both classes have the same a priori probability of 0.5.

The following example shows how weights are updated depending on the error level in the second step, $b = 2$. For this, two values are selected near the limits of the ε_b domain (0 and 0.5). If $\varepsilon_b = 0.499$, then $\alpha_b = 0.004$ and the new weight for the second step is $w_2(i) = 1/n \exp(0.004\mathbf{I}(C_1(\mathbf{x}_i) \neq y_i))$. Therefore, if the ith observation is wrongly classified, its weight $w_2(i) = 1/n \cdot 1.004$, whereas if it is correctly classified its weight would remain constant but it finally will be reduced due to the normalization process. Assuming that $\varepsilon_b = 0.001$, then $\alpha_b = 6.907$ and the weight of a wrongly classified example will be $w_2(i) = (1/n)999$, whereas the weight of the correctly classified examples will be reduced in the normalization.

Table 3.1 displays the weights for the second iteration after the normalization process for a set of 1000 observations. It shows how the weight of the wrongly classified observations increases and the weight of the correctly classified ones decreases, forcing the single classifier built into the following iteration to focus on the hardest examples. In addition, the differences when the weights are updated are greater when the error of the single classifier is small since more importance is given to the few mistakes mentioned when the classifier achieves a high level of accuracy. The alpha constant can therefore be interpreted as a learning rate which is calculated as a function of the error made on each iteration. This constant is also used in the final decision rule, giving more importance to the individual classifiers that made a smaller error. Recent works have focused on control of the emphasis weighting function used to train each new learner of the ensemble (Ahachad et al., 2017).

Table 3.1 Example of the weight updating process in AdaBoost.

n	Initial weight	Error	Alpha	Wrongly classified	Weight (2)	Weight (2) normalized
1000	0.001	0.499	0.004	Yes	0.001004	0.001002004
1000	0.001	0.499	0.004	No	0.001000	0.000998004
1000	0.001	0.001	6.907	Yes	0.999000	0.500000000
1000	0.001	0.001	6.907	No	0.001000	0.000500501

This process is repeated in every step for $b = 1, 2, 3, \dots, B$. Finally, the ensemble classifier is built as a linear combination of the single classifiers weighted by the corresponding constant α_b:

$$C_f(\mathbf{x}_i) = sign\left(\sum_{b=1}^{B} \alpha_b C_b(\mathbf{x}_i)\right) \tag{3.6}$$

Algorithm 3.2 summarizes the AdaBoost pseudocode.

Algorithm 3.2 AdaBoost

1. Start with $w_1(i) = 1/n,\ i = 1, 2, \dots, n$.
2. Repeat for $b = 1, 2, \dots, B$.
   a Fit the classifier $C_b(\mathbf{x}_i) \in \{-1, 1\}$ using the weights $w_b(i)$ in $\mathbf{T}_b$.
   b Compute: $\varepsilon_b = \sum_{i=1}^{n} w_b(i)\mathbf{I}(C_b(\mathbf{x}_i) \neq y_i)$ y $\alpha_b = ln((1 - \varepsilon_b)/\varepsilon_b)$.
   c Update the weights $w_{b+1}(i) = w_b(i)\exp\left(\alpha_b \mathbf{I}(C_b(\mathbf{x}_i) \neq y_i)\right)$ and normalize them.
3. Output the final classifier:

$$C_f(\mathbf{x}_i) = sign\left(\sum_{b=1}^{B} \alpha_b C_b(\mathbf{x}_i)\right)$$

AdaBoost can be applied in two different ways, using resampling or reweighting. In the version that uses resampling, we obtain the set of data $\mathbf{T}_b$ for the bth iteration, by means of a subsample bootstrap extracted with replacement, using as probabilities for the extraction the weights of the different observations in that iteration. In the version that uses reweighting, the classifier $C_b(\mathbf{x}_i)$ takes directly into account the weights of the examples. There is no strong evidence in favor of either method versus the other (Breiman, 1998; Freund and Schapire, 1997, 1998).

The most commonly used method to build the basic boosting classifiers are the classification trees and these can discriminate between more than two classes. It therefore seems appropriate to use boosting algorithms able to handle the multiclass case. Among the versions of boosting algorithms for multiclass classification problems (Mukherjee and Schapire, 2011), two of the most simple and natural extensions of AdaBoost have been chosen here, AdaBoost.M1 and SAMME.

The first algorithm, AdaBoost.M1, with only slight modifications, adapts AdaBoost to the general case of k classes. The basic classifier constructed during the bth iteration in step 2a of Algorithm 3.2, for example a classification tree, is able to distinguish between more than two classes, since in this case it is not forced to distinguish only between $\{-1, 1\}$. In step 2b, the classifier error of that iteration is calculated by adding the weights of the misclassified examples, that is, those in which its actual class does not match the class predicted by the basic classifier of that iteration, whichever of the possible k classes. Therefore, in this case the training set is given by $\mathbf{T}_n = \{(\mathbf{x}_1, y_1), \dots, (\mathbf{x}_n, y_n)\}$, where the y_i values are $\{1, \dots, k\}$.

Algorithm 3.3 AdaBoost.M1

1. Start with $w_1(i) = 1/n,\ i = 1, 2, \ldots, n$.
2. Repeat for $b = 1, 2, \ldots, B$.
   a Fit the classifier $C_b(\mathbf{x}_i) \in \{1, 2, \ldots, k\}$ using the weights $w_b(i)$ in $\mathbf{T}_b$.
   b Compute: $\varepsilon_b = \sum_{i=1}^{n} w_b(i)\mathbf{I}(C_b(\mathbf{x}_i) \neq y_i)$ y $\alpha_b = ln((1-\varepsilon_b)/\varepsilon_b)$.
   c Update the weights $w_{b+1}(i) = w_b(i)\exp\left(\alpha_b \mathbf{I}(C_b(\mathbf{x}_i) \neq y_i)\right)$ and normalize them.
3. Output the final classifier:

$$C_f(\mathbf{x}_i) = \arg\max_{y_j} \sum_{b=1}^{B} \alpha_b \mathbf{I}(C_b(\mathbf{x}_i) = y_j)$$

The final classifier AdaBoost uses the sign of the weighted sum of the results of the individual classifiers. Now the final classifier (AdaBoost.M1), upon a new observation, will collect for each class the weighted sum of votes that each class has received from the B basic classifiers. Therefore, the predicted class will be the one that has obtained a greater value in the weighted vote. This expression is valid for both the dichotomous case and the general case. In addition, in the dichotomous case the expression 3.7 is equivalent to equation 3.6:

$$C_f(\mathbf{x}_i) = \arg\max_{y_j} \sum_{b=1}^{B} \alpha_b \mathbf{I}(C_b(\mathbf{x}_i) = y_j) = \arg\max_{y_j} \sum_{b:C_b(\mathbf{x}_i)=y_j} \alpha_b \qquad (3.7)$$

As we did with AdaBoost, we can see an example of how the weights are updated depending on the error committed for the second iteration, $b = 2$. For this, we have chosen two values close to the extremes of range of ε_b, 0 and 0.5. If $\varepsilon_b = 0.499$, then $\alpha_b = 0.002$ and the new weight for the second iteration will be $w_2(i) = (1/n)\exp(-0.0002\mathbf{I}(C_2(\mathbf{x}_i) \neq y_i))$. Then, if the ith observation is incorrectly classified its weight will be $w_2(i) = (1/n)1.002$, whereas if it has been classified correctly, its weight in principle is not modified, but will decrease when normalizing. The goal of normalization is that the sum of all weights is one. If we now consider that $\varepsilon_b = 0.001$, then $\alpha_b = 3.453$ and the weight of a misclassified observation will have $w_2(i) = (1/n)31.6$, while the weights of correctly classified observations will be reduced later in standardization. Table 3.2 shows how the weights for the second iteration would remain once normalized for a set of 1000 observations.

Finally, the **S**tagewise **A**dditive **M**odeling algorithm using a **M**ulti-class **E**xponential loss function (SAMME) (Zhu et al., 2009) also generalizes AdaBoost for more than two classes. Previously, Eibl and Pfeiffer (2002) presented AdaBoost.M1W, which is similar to SAMME. It should be noted that the only difference between this algorithm and AdaBoost.M1 is in the form of calculating the alpha constant, since in this case the number of classes is taken into account. Due to this modification, the SAMME algorithm only needs $1 - e_b > 1/k$ so that the alpha constant is positive and the updating of the weights follows the correct direction. That is, the accuracy of each weak classifier should

Table 3.2 Example of the weight updating process in AdaBoost.M1.

n	Initial weight	Error	Alpha	Wrongly classified	Weight (2)	Weight (2) normalized
1000	0.001	0.499	0.002	Yes	0.001002002	0.001001002
1000	0.001	0.499	0.002	No	0.001000000	0.000999002
1000	0.001	0.001	3.452	Yes	0.031606961	0.030668298
1000	0.001	0.001	3.452	No	0.001000000	0.000970302

be better than the default rule instead of 1/2, which is a suitable requirement when there are only two classes but too demanding for the multiclass case.

To conclude this section, we collect some other works related to generalization for more than two classes. Eibl and Pfeiffer (2005) present two algorithms, GrPloss and BoostMA, where the basic classifiers must minimize the pseudo-loss error and the error corrected by the default rule (maxlabel). Li (2006) proposes a new multiclass boosting algorithm that modifies the coding matrix as a function of the learning ability of the basic classifier. Rodriguez (2009) compares some of the versions of boosting for more than two classes. Finally, Mukherjee and Schapire (2010) discuss the correct requirements for weak classifiers in boosting for the case of more than two classes.

3.4.1 AdaBoost Training Error

Freund and Schapire showed that by increasing the number of iterations B, the training or apparent error of the combined AdaBoost classifier tends to zero at an exponential rate. The weight updating rule in step 2 of Algorithm 3.2 implies that

$$\begin{aligned}\sum_{i=1}^{n} w_{b+1}(i) &= \sum_{i=1}^{n} w_b(i)\exp(\alpha_b \mathbf{I}(C_b(\mathbf{x}_i) \neq y_i)) \\ &= \sum_{i=1}^{n} w_b(i)\left(\frac{1-\varepsilon_b}{\varepsilon_b}\right)^{\mathbf{I}(C_b(\mathbf{x}_i)\neq y_i)} \leq \\ &\langle\langle \alpha^r \leq 1-(1-\alpha)r \text{ for } \alpha \geq 0 \text{ and } r \in [0,1] \text{ therefore}\rangle\rangle \\ &\leq \sum_{i=1}^{n} w_b(i)\left(1-\left(1-\frac{1-\varepsilon_b}{\varepsilon_b}\right)\mathbf{I}(C_b(\mathbf{x}_i) \neq y_i)\right) \\ &= \left(\sum_{i=1}^{n} w_b(i)\right)\left(1-\left(1-\frac{1-\varepsilon_b}{\varepsilon_b}\right)\varepsilon_b\right)\end{aligned} \tag{3.8}$$

combining this inequality on $b = 1, 2, 3, \ldots, B$, we have that

$$\sum_{i=1}^{n} w_{B+1}(i) = \prod_{b=1}^{B}\left(1-\left(1-\frac{1-\varepsilon_b}{\varepsilon_b}\right)\varepsilon_b\right) \tag{3.9}$$

On the other hand, the final decision rule will fail on the ith observation if

$$\prod_{b=1}^{B}\left(\frac{1-\varepsilon_b}{\varepsilon_b}\right)^{\mathbf{I}(C_b(\mathbf{x}_i)\neq y_i)} \geq \prod_{b=1}^{B}\left(\frac{1-\varepsilon_b}{\varepsilon_b}\right)^{\frac{1}{2}} \tag{3.10}$$

and the final weight of any instance is

$$w_{B+1}(i) = w_1(i) \prod_{b=1}^{B} \left(\frac{1-\varepsilon_b}{\varepsilon_b} \right)^{\mathbf{I}(C_b(\mathbf{x}_i) \neq y_i)} \tag{3.11}$$

Combining equations 3.10 and 3.11, a lower bound for the sum of the final weights can be set as the sum of the final weights of the observations where the final decision rule is wrong:

$$\begin{aligned} \sum_{i=1}^{n} w_{B+1}(i) \geq \sum_{i:C_f(\mathbf{x}_i) \neq y_i} w_{B+1}(i) \geq \\ \geq \left(\sum_{i:C_f(\mathbf{x}_i) \neq y_i} w_1(i) \right) \left(\prod_{b=1}^{B} \frac{1-\varepsilon_b}{\varepsilon_b} \right)^{\frac{1}{2}} = \varepsilon_A \left(\prod_{b=1}^{B} \frac{1-\varepsilon_b}{\varepsilon_b} \right)^{\frac{1}{2}} \end{aligned} \tag{3.12}$$

where ε_A is the final classifier error in the learning set, or apparent error. Combining the results of equations 3.9 and 3.12, we have:

$$\begin{aligned} \varepsilon_A \leq \prod_{b=1}^{B} \frac{1 - \left(1 - \frac{1-\varepsilon_b}{\varepsilon_b}\right) \varepsilon_b}{\sqrt{\frac{1-\varepsilon_b}{\varepsilon_b}}} = \prod_{b=1}^{B} \frac{1-(2\varepsilon_b - 1)}{\sqrt{\frac{1-\varepsilon_b}{\varepsilon_b}}} = \\ = \prod_{b=1}^{B} \frac{2(1-\varepsilon_b)}{\sqrt{\frac{1-\varepsilon_b}{\varepsilon_b}}} = 2^B \prod_{b=1}^{B} \sqrt{\varepsilon_b(1-\varepsilon_b)} \end{aligned} \tag{3.13}$$

where as $\varepsilon_b < 0.5$, then $\varepsilon_b(1-\varepsilon_b) < 0.25$ and, therefore, this expression shows that as the number of iterations (B) increases, the apparent error tends to zero at an exponential rate.

3.4.2 AdaBoost and the Margin Theory

The concept of margin arises with the attempt of Shapire et al. to explain the empirically proven fact that boosting does not suffer from overfitting even when a large number of classifiers are combined, i.e. the test set error continues decreasing with the number of combined classifiers even after the training error becomes null. This led to a renewed interest in the search for possible explanations and gave rise to the so-called margin theory (Schapire et al., 1998).

The concept of the margin comes from the theory of statistical learning and is related to the Vapnik–Chervonenkis dimension (commonly referred to as the VC dimension). Simplyfing, the VC dimension can be said to provide an upper bound on the accuracy of classification models. Although the boundary is broad, it has been shown to be an important theoretical tool in pattern recognition and machine learning. The classification method known as the support vector machine (SVM) is based on the statistical learning theory and, more specifically, on the idea of maximizing margins. Intuitively, the margin of an observation is related to the certainty or confidence of its classification. Observations for which the assigned class is correct and has a high degree of confidence will have positive and large margins. Examples with uncertain or doubtful classification should have small margins. A small margin is a symptom of instability in the assigned class, i.e. the example can be assigned to different classes by similar classifiers (Kuncheva, 2004).

For k classes, the margin of an example $\mathbf{x}_i$ is calculated using the votes for every class j in the final ensemble, which is known as the degree of support of the different classes or posterior probabilities $\mu_j(\mathbf{x}_i)$, $j = 1, 2, \dots, k$ as

$$m(\mathbf{x}_i) = \mu_{y_i}(\mathbf{x}_i) - \max_{j \neq y_i} \mu_j(\mathbf{x}_i) \tag{3.14}$$

where y_i is the correct class of $\mathbf{x}_i$ and $\sum_{j=1}^{k} \mu_j(\mathbf{x}_i) = 1$. All the wrongly classified examples will therefore have negative margins and the correctly classified ones will have positive margins.

Correctly classified observations with a high degree of confidence will have margins close to one. On the other hand, examples with an uncertain classification will have small margins, that is to say, margins close to zero. Since a small margin is an instability symptom in the assigned class, the same example could be assigned to different classes by similar classifiers.

For visualization purposes, Schapire et al. (1998) use margin distribution graphs showing the cumulative distribution of the margins for a given data set. The x axis is the margin (m) and the y axis is the number of points where the margin is less than or equal to m. Ideally, all points should be classified correctly so that all the margins are positive. If all the points have been classified correctly and with the maximum possible certainty, the cumulative graph will be a single vertical line at $m = 1$.

Boosting the margins, we can achieve classifiers with higher confidence. Schapire et al. (1998) demonstrated upper limits for margin-dependent test errors. The main theoretical result for the dichotomous problem is given in the following theorem.

Theorem 3.1 Let $\mathbf{H}$ be a finite space of basic classifiers.[1] For $\delta > 0$ and $\theta > 0$, with probability at least $1 - \delta$ on the random selection of training set $\mathbf{T}$, any combination of classifiers $C_f = \{C_1, C_2, \dots, C_B\} \subseteq \mathbf{H}$ using the weighted mean satisfies

$$P(error) \leq P(training\ margin \leq \theta) + O\left(\frac{1}{\sqrt{n}}\left(\frac{\ln n \ln |\mathbf{H}|}{\theta^2} + \ln(1+\delta)\right)^{1/2}\right) \tag{3.15}$$

where $P(error)$ is the probability that the combination fails by classifying $\mathbf{x}_i \in \mathbb{R}^n$ and $P(training\ margin \leq \theta)$ is the probability that the margin of an example drawn at random from the training set does not exceed θ. $|\mathbf{H}|$ is the cardinality of $\mathbf{H}$. For the more general case of finite or infinite $\mathbf{H}$ with VC dimension d, the following limit is used, if $1 \leq d \leq n$

$$P(error) \leq P(training\ margin \leq \theta) + O\left(\frac{1}{\sqrt{n}}\left(\frac{d \ln^2(n/d)}{\theta^2} + \ln(1+\delta)\right)^{1/2}\right) \tag{3.16}$$

1 A finite space of basic classifiers is, for example, the set of all classification trees of a given size on a set of discrete variables. For example, the set of all decision stumps on a set of p dichotomous variables contains p elements (classifiers), one for each variable.

None of these limits depends on the number of classifiers in the ensemble. Although the limits are quite loose, they show that large margins lead to smaller upper limits in the test error.

3.4.3 Other Boosting Versions

The AdaBoost algorithm described in section 3.4 trains the classifier several times in the same training set, at every step it focuses its attention on the examples that have been misclassified in the previous step. AdaBoost is limited to the dichotomous case and uses basic classifiers whose output can only take values -1 and 1, so it is also known as discrete AdaBoost, from the work of Friedman et al. (2000). Since its appearance in 1996, many modifications have been proposed to improve the boosting algorithms, but in this work only some of them are collected. See Schapire and Freund (2012) or Wu and Nagahashi (2015) for a more detailed compilation of boosting variants.

A generalization of the discrete AdaBoost was proposed in Freund and Schapire (1996) and explained in greater depth in Schapire and Singer (1999). This generalization uses predictions with real values, which also express the degree of confidence, instead of the values $\{-1, 1\}$ using AdaBoost. In this case, the basic classifiers assign a real value between -1 and 1. Thus, the sign of $C_b(x)$ reports the predicted class and the absolute value, $|C_b(x)|$, gives a measure of confidence in prediction. The actual value of this contribution is combined with the previous ones using, as in the original algorithm, a coefficient α_b, although this time it is calculated in a slightly different way.

Friedman et al. (2000) presented a version called **Real Adaboost**, where the basic classifiers estimate the probability of belonging to a certain class $p_b(x) = P_w(Y = 1/X = x) \in [0, 1]$, where the subscript w indicates that this probability is calculated on the weighted set using the weights $w_b(i)$ corresponding to each iteration. The contribution to the final classifier is half the logit transformation of this estimated probability.

Algorithm 3.4 Real AdaBoost

1. Start with $w_1(i) = 1/n,\ i = 1, 2, ..., n$.
2. Repeat for $b = 1, 2, \ldots, B$.
   a Construct a classifier to obtain an estimate of the probabilities of the classes $p_b(x) = \hat{P}_w(y = 1/x) \in [0, 1]$ using weights $w_b(i)$ in $\mathbf{T}_b$.
   b Compute: $f_b(x) = 1/2 \ln \left(p_b(x)/(1 - p_b(x))\right) \in \mathbb{R}$.
   c Update the weights $w_{b+1}(i) = w_b(i) \exp(-y_b f_b(x_i))$ and normalize them.
3. Output the final classifier:

$$C_f(x) = sign\left(\sum_{b=1}^{B} f_b(x)\right)$$

As we have seen in section 3.4, the AdaBoost algorithm minimizes an exponential loss function. However, several authors argue that for classification, instead of an exponential loss function, the choice of the logarithm of binomial likelihood as a loss function becomes more natural. Based on this function Friedman et al. (2000) proposed a new algorithm called **LogitBoost**. This new algorithm maintains the same philosophy of the previous boosting algorithms, to train in successive occasions to the classifier, forcing it to focus on the difficult examples, but the way of carrying it out is different.

In each iteration, a function $f_b(x_i)$ is created that fits as closely as possible to z_i^b, which is the objective response with which the bth iteration is used for the ith observation. This answer is calculated according to the true class of that observation, which in this case will be $\{0, 1\}$, to the probability that has been assigned to class 1 in the previous iteration, and to the weight in that iteration:

$$z_i^b = \frac{y_i - p_{b-1}(x_i)}{w_i^b} \tag{3.17}$$

where $p_{b-1}(x_i) = P_w(Y = 1|X = x_i)$ and $w_b(i) = p_{b-1}(x_i) \cdot (1 - p_{b-1}(x_i))$.

$\mathrm{F}_b(x_i)$ is an aggregation function that is updated at each iteration as $\mathrm{F}_b(x_i) = \mathrm{F}_{b-1}(x_i) + 0.5f_b(x_i)$. Next, this function is used to calculate the probability that the ith observation belongs to class 1 in iteration b:

$$p_b(x_i) = \frac{\exp(\mathrm{F}_b(x_i))}{\exp(\mathrm{F}_b(x_i)) + \exp(-\mathrm{F}_b(x_i))} = \frac{1}{1 + \exp(-2\mathrm{F}_b(x_i))} \tag{3.18}$$

To better understand this algorithm, each aggregation function, $\mathrm{F}_b(x_i)$, can be considered as an estimate of half of the natural logarithm of the probabilities ratio of the classes:

$$\mathrm{F}(x) = \frac{1}{2} \ln\left(\frac{p(x)}{1 - p(x)}\right) \tag{3.19}$$

Therefore, **Logitboost** adjusts an additive logistic regression model by sequential optimization of the logarithm of binomial likelihood (for details see Friedman et al. (2000)).

Algorithm 3.5 LogitBoost

1. Start with the aggregation function $F_0(x) \equiv 0$ and probabilities $p_0(x) = 1/2$, where $p(x)$ is the abbreviated form of $\hat{P}[Y = 1|X = x]$.

2. Repeat for $b = 1, 2, \ldots, B$.

   a Compute the work response and weights for $i = 1, 2, \ldots, n$:

   $$w_b(i) = p_{b-1}(x_i)(1 - p_{b-1}(x_i)) \ : \ z_i^b = \frac{y_i - p_{b-1}(x_i)}{w_i^b}$$

   b Adjust $f_b(x_i)$by weighted least squares:

   $$f_b(x_i) = \arg\min_f \sum_{i=1}^{n} w_b(i)(z_i^b - f_b(x_i))^2$$

   c Update the aggregation function and calculate the new probabilities:

   $$\mathrm{F}_b(x_i) = \mathrm{F}_{b-1}(x_i) + 0.5f_b(x_i)$$

   $$p_b(x_i) = (1 + \exp(-2\mathrm{F}_b(x_i)))^{-1}$$

3. Output the final classifier:

   $$C_f(x_i) = sign(\mathrm{F}_b(x_i))$$

A very useful property of this method is that it directly produces estimates of the probabilities $\hat{P}[Y = 1/X = x]$. This is very important for building classifiers when the error costs in the classification are not the same. In addition, it allows classifiers to be built with the option of not assigning any class or assigning the label "doubt" or "no class" for certain observations. An important advantage of **Logitboost** compared to methods such as neural networks is that it works well without the need for fine tuning and without performing a sophisticated nonlinear optimization, although this advantage is general for boosting algorithms.

Given the definition of the weights $w(x)$ in those examples where $p(x)$ is close to 0 or 1, $w(x)$ will become very small. This can cause problems in the calculation of $z(x_i)$, so Friedman et al. (2000) advise taking the following precautions.

If $y = 1$, then calculate $z = (y - p)/p(1 - p)$ as $1/p$. Since this quotient can become very large if p is small, its value must be limited by a certain amount, which can be represented by *zmax*. The specific value chosen for *zmax* is not crucial and Friedman (2000) states that values between 2 and 4 work well. On the other hand, if $y = 0$, calculate z as $-1/(1 - p)$ with the lower limit of $-zmax$.

Leo Breiman, author of the bagging method (section 3.3), showed a great interest in the boosting method that Freund and Schapire proposed almost simultaneously to bagging and with better results than this one. He considered that the fundamental of this technique is the use of adaptive resampling and combination, and for that reason calls them **Arcing algorithms.**[2] The main difference with the bagging method is training the basic classifiers in samples generated from probability distributions that change according to the mistakes made, instead of remaining constant, as in bagging. This implies that the basic classifiers used in bagging are independent, whereas in the case of boosting each classifier depends on the previous ones.

Algorithm 3.6 Arcing

1. Start with $w_1(i) = 1/n,\ i = 1, 2, ..., n$.
2. Repeat for $b = 1, 2, \dots, B$.
   a Fit the classifier $C_b(\mathbf{x}_i)$ using weights $w_b(i)$ in $\mathbf{T}_b$.
   b Classify the examples of $\mathbf{T}$ in using $C_b(x)$ and compute $m_b(x_i)$, which is the number of errors made by the first b classifiers in the ith observation.
   c Update the weights $w_{b+1}(i) = \frac{1+m_b(x_i)^4}{\sum_{i=1}^{n}\left(1 + m_b(x_i)^4\right)}$.
3. Output the final classifier:

$$C_f(x) = sign\left(\sum_{b=1}^{B} C_b(x)\right)$$

2 Arcing is an acronym of Adaptive Resampling and Combining. Breiman refers to AdaBoost as arc-fs, in honor of Freund and Schapire, and arc-x4 to the modification that he proposes, but in this work when we use arcing it will refer to arc-x4.

Breiman (1998) explains that after testing AdaBoost he began to suspect that its success was not in its concrete form but in the adaptive resampling that it performs, where the weights of those examples that are more frequently erroneously classified are increased. To prove this, he tested three simple ways to update weights or probabilities. In all of them, the update is performed as a function of the value $1 + m_i^h$, where m_i is the number of times the ith observation has been misclassified by the basic classifiers constructed up to that iteration. Breiman used the values $h = 1, 2, 4$ and the latter gave the best results. Another difference of arcing with respect to boosting is that it does not use weights in the final combination rule, which also means a greater simplicity of the algorithm by not having to calculate and save those weights.

3.4.4 Comparing Bagging and Boosting

Both methods construct the basic classifiers in modified versions of the training set and then combine them into the final classification rule. However, they differ in how to obtain the basic classifiers. Specifically, the three main differences are:

1. All observations are used in each step of boosting (no bootstrapping), at least in its initial version in Freund and Schapire (1996).
2. The final decision rule of bagging does not weigh the basic classifiers differently per their accuracy, whereas in boosting it does.
3. Finally, but at the same time most importantly, in boosting each classifier depends on the previous ones, while in bagging the classifiers are independent.

3.5 Random Forests

As noted in section 2.4, Krogh and Vedelsby (1995) argue that the ideal combination is to add basic high accuracy classifiers that disagree as often as possible, that is, basic classifiers that have a low error rate and are scarcely related to each other.

Following this idea, some methods have tried to introduce randomness in the process of construction of the basic classifiers. In this way, decreasing the relationship between them is desired, to increase the accuracy of the combined classifier. The randomness can be introduced either in the learning algorithm of the basic classifier or in the training set used to build it.

When using classification trees as base classifiers, there are several methods that generate random vectors that determine the growth of each tree in the combination. The first of these methods is the already studied bagging (Breiman, 1996a), where each tree is developed on a bootstrap replica of the original training set, that is, a random selection is made with replacement of the observations in the training set.

Another example is the method of random selection of the split (Dietterich, 2000b), where, at each node, the split is selected randomly among the k best possible splits, that is, the information gain of all possible splits is first calculated. They are then ranked and, among the best k, one is chosen at random. The value of k is fixed previously and is the same in all the nodes. Third, Ho (1998) proposed the method of random subspace, which randomly selects a group of variables to be used in the construction of each tree. Finally, Breiman (1999) proposed a method that generates new training sets by randomizing the outputs in the original training set.

The common element in all these methods is that for the ith tree a random vector $\mathbf{\Theta}_i$, independent of the random vectors used to construct the previous trees, $(\mathbf{\Theta}_1, \mathbf{\Theta}_2, \dots, \mathbf{\Theta}_{i-1})$, but the same distribution, is generated. Therefore, the ith tree is constructed using the training set and $\mathbf{\Theta}_i$, obtaining a classifier $C(\mathbf{x}_i, \mathbf{\Theta}_i)$, where $\mathbf{x}_i$ is the characteristics vector. For example, in bagging, the random vector $\mathbf{\Theta}$ is generated by drawing, with replacement, a sample of the same size as the original training set, whereas in the random selection of the split, $\mathbf{\Theta}$ consists of a vector of integers randomly and independently selected between 1 and k.

After generating a sufficiently large number of basic trees, their votes are collected to see which is the majority class. Breiman (1999) calls these procedures random forests.

Specifically, Breiman (1999) defines a **random forest** as a classifier consisting of a set of classifiers with tree structure $\{C(\mathbf{x}_i, \mathbf{\Theta}_i); i = 1, 2, \dots\}$, where $\{\mathbf{\Theta}_i\}$ are independent and identically distributed random vectors and each tree collects a unitary vote for the majority class at a given observation $\mathbf{x}_i$.

From this definition, two aspects that differentiate the random forests from boosting should be emphasized. First, the vectors $\mathbf{\Theta}$ are independent and identically distributed, while in boosting the vector of weights in an iteration depends on the behavior in the previous iterations. Second, in a random forest each tree provides a unitary vote for the determination of the majority class, but in boosting the vote of each tree is weighted by the constant α, which is calculated from the error of that tree.

Breiman proposes two ways of constructing the random forests, both with the same objective, to obtain accurate basic classifiers and, simultaneously, make them the least related to each other that they can be. The first option constructs random forests by random selection of input variables and the second constructs them by random linear combinations of these variables.

Random forests with random selection of the input variables imply the simultaneous use of bagging and the random selection of variables. From the original data set, bootstrap replicas are performed, selecting with replacements the same number of elements as in the original data set. Next, a tree is constructed for each of the new training sets using a random selection of variables. That is, at each node, a small group of characteristics is randomly selected, among which the most appropriate one should be chosen to perform the split in that node. The tree is completely developed without pruning. The size of the selected group of variables, F, must be set previously. Breiman (1999) tested the values $F = 1$ and the first integer smaller than $\log_2(p + 1)$, where p is the number of input variables that characterize each observation. Later, the same author advised setting the value F in the square root of p, although according to him the procedure is not very sensitive to the size set.

Algorithm 3.7 Random forest

1. Set the number of trees to grow.
2. For each tree:
   a Draw a random subset $\mathbf{T}_k$ of the training set $\mathbf{T}$ (N cases with replacement) to train each tree. The elements in $\mathbf{T}$ but not in $\mathbf{T}_k$ are called out-of-bag (oob).
   b Set F (number of variables to do a split) $<< p$ (number of input variables) and choose the best split among the F randomly selected variables for each node in each tree.
   c Grow the tree to maximum size.
   d Use oob training data to estimate error and variable importance.
3. Assign a class to new data as the majority vote among all the trees.
4. Estimate the classification accuracy (or error) for the random forest and the importance measure for each input variable.

In the experiments performed in 20 sets, frequently used in machine learning, Breiman is surprised by the good results obtained by selecting a single variable since they are, in general, only slightly inferior to those achieved by selecting a group, and even on occasions higher than these.

The random selection of variables streamlines the process, since the number of variables for which the information gain at each node has to be calculated is reduced. Therefore, the construction of a random forest in this way will be faster than the construction of bagging. In addition, it is important to highlight the difference between the random selection of variables and the random selection of the split. In the first, a small group of variables is chosen at random and among them the one that produces the best split is selected. In the random selection of the split, the gain of information produced by all the variables in that node is calculated, and later a split of the best k is chosen at random. That is, in the random selection of variables, the information gain must be calculated only for those randomly selected variables, while in the random selection of the split, the information gain is calculated for all variables. For data sets with many variables the calculation time difference may be important. In the big data era, this time saving could be an important advantage.

If the available number of features is small, the random selection of variables by making F a considerable portion of p can achieve an improvement in the accuracy of the basic classifiers, but will increase the relation between them. Another possibility is to generate new variables starting from the random linear combination of a group of characteristics. A variable is generated by specifying the number of variables to be combined, L. At each node, L variables are randomly selected and joined using coefficients that are obtained from a uniform distribution in the interval $[-1, 1]$. In the same way, F linear combinations are generated and, subsequently, the search for the best split between them is performed. This is the random forest procedure with random linear combinations of the variables. Breiman (1999) uses $L = 3$ and $F = 2$ or 8, being the value of F

chosen as a function of the estimation of the error. The value $L = 3$ allows the number of possible combinations to be sufficient so that high values of F achieve a high accuracy of the basic classifiers, but without causing a very strong relation between them.

Summarizing, it can be said that random forests consist of using a random selection of variables, or random combinations of them, at each node during the construction of the basic classifiers. According to Breiman, this type of procedure has some desirable properties, for example its accuracy is similar to that of AdaBoost, it is built faster than bagging and boosting, and it behaves well when the discriminatory power is widely distributed among many variables, which is a difficulty for many individual classifiers. In addition, it is relatively robust against outliers and the presence of noise.

Random forests do not concentrate weights on any subset of the examples, and therefore the effect of noise is less, as opposed to other methods, such as AdaBoost, that raise the weight of the misclassified examples. Noisy observations, i.e. those whose class label is incorrect, will be persistently misclassified, therefore AdaBoost will focus erroneously on those noisy observations and will be distorted.

There are several studies that compare this type of method, among others we can mention the one by Pino-Mejias et al. (2008) that compares random forests with bagging and boosting using classification trees, logistic regression, multilayer perceptron, and SVM as base classifiers, all with a medical application (fragile X syndrome). Bagging with SVM gets slightly better results. In addition, some works try to solve these weaknesses of boosting, thus Sun et al. (2007) proposed a modification of boosting with cost sensitivity for data with imbalanced classes and Warmuth et al. (2008) proposed regularizing boosting for data sets with noise.

Out-of-bag Estimations

Tibshirani (1996a) proposed a bootstrap estimate of a bagging classifier error. Breiman (1999) calls this estimation an out-of-bag estimate or outside the sample. The objective is to take advantage of the observations of the training set that do not become part of each of the bootstrap replicas used to build the basic classifiers. Since these observations do not contribute to the formation of that classifier, they can be a good set in which to prove it.

B bootstrap replicas ($\mathbf{T}_b$) are built from the original set ($\mathbf{T}_n$) and in each a basic classifier ($C_b(\mathbf{x}_i)$) is trained. For each observation $\mathbf{x}_i$ of $\mathbf{T}_n$, we add the votes of the basic classifiers where $\mathbf{x}_i \notin \mathbf{T}_b$, that is, those basic classifiers that have been constructed without including $\mathbf{x}_i$ in their training set. This combined classifier is the out-of-bag or out-of-sample classifier. The out-of-bag estimate for the generalization error is the error rate of the out-of-bag classifier in the set $\mathbf{T}_n$.

In each bootstrap replica, approximately 37% of the observations remain outside. Therefore, the out-of-bag estimates are based on combining about one-third of the classifiers used in the current main ensemble. Since the error rate decreases as the number of combined classifiers increases, the out-of-bag estimate will tend to overestimate the error rate in the current iteration. To obtain unbiased out-of-bag estimates, it is necessary to continue the process beyond the point where the test error rate converges. Therefore, unlike cross-validation, where there is an unknown amount of bias, out-of-bag estimates are unbiased.

As Wolpert and Macready (1999) point out, cross-validation estimations in ensemble classifiers using numerous bootstrap replicas require large computational efforts. On the other hand, the out-of-bag estimates are efficient in this sense because they can be calculated at the same time as the combined classifier is constructed with a very small additional effort. Breiman (1996b) argues that these estimates, moreover, are close to the optimum, according to the conclusions he draws from his experiments.

4

Classification with Individual and Ensemble Trees in R

Esteban Alfaro, Matías Gámez, and Noelia García

4.1 Introduction

To apply the methods collected in this manual the R programming environment is used (R Development Core Team, 2017), consisting of a set of packages or libraries for data manipulation, calculation, and graphics. Among other features, it has a well-developed and effective programming language (R language), it is considered by the researcher community to be one of the most powerful tools to implement statistical methods or, in a more general sense, to do any data analysis, it is an open source programming language with a wide and very active community contributing to its development, and, of course, it is freely distributed, which ensures that it can be used not only by supported researchers but by students all over the world.

Specifically, for practical applications version R 3.4.0 (April 2017) has been used, although updates of this program occur very frequently, therefore there is no doubt that by the time this work sees the light there will have been more updated versions. Some statistical techniques are included in the base package of R and others are available in the form of libraries (packages) on the Internet in CRAN (http://www.r-project.org). These libraries collect a set of functions that have a series of parameters that can be adjusted and whose meaning can be found through the help function that each library has.

Among the libraries available in CRAN is **adabag** (Alfaro et al., 2013), which implements the AdaBoost.M1, AdaBoost-SAMME, and bagging algorithms with CART trees (Breiman et al., 1984) as base classifiers, constructed using the **rpart** library (Therneau and Atkinson, 2017). These ensemble methods can use any base classifier, so it should be clarified that in the remainder of this work when talking about boosting or bagging we mean boosted or bagged trees.

The **adabag** package consists of a total of 13 functions, three for each method, three functions for plots, and the autoprune, evolerror, margins, and MarginOrderedPruning.Bagging. The three functions for each method are to build the boosting (or bagging) classifier and classify the samples in the training set, to predict the class of new samples using the previously trained ensemble, and to estimate by cross-validation the accuracy of these classifiers in a data set. The function autoprune prunes automatically an object of class `rpart`. The error evolution and the margin in the class prediction

Ensemble Classification Methods with Applications in R, First Edition.
Edited by Esteban Alfaro, Matías Gámez, and Noelia García.

for each observation can be calculated. Finally, MarginOrderedPruning.Bagging applies margin-based ordered aggregation for bagging pruning (Guo and Boukir, 2013).

Since the first version of this library, in 2006, it has been used in several classification tasks from very different scientific fields, such as economics and finance (Alfaro et al., 2008a,b; Chrzanowska et al., 2009a; De Bock and Van den Poel, 2011), automated content analysis (Stewart and Zhukov, 2009), interpreting signals out of control in quality control (Alfaro et al., 2015), and advances in machine learning (De Bock et al., 2010; Krempl and Hofer, 2008). Moreover, **adabag** is used at least in three R manuals (Maindonald and Braun, 2010; Torgo, 2010; Alfaro et al., 2014).

There are four packages with reverse dependencies to **adabag**. The package **rminer** (Cortez, 2016) reverse imports **adabag** and facilitates the use of data mining algorithms in classification and regression (including time series forecasting) tasks. Reverse suggestions of **adabag** include **crtests** (van der Spoel, 2016), which provides wrapper functions for running classification and regression tests using different machine learning techniques, such as random forests and decision trees. The **mlr** package (Bischl et al., 2016) is an interface to many classification and regression techniques, including machine-readable parameter descriptions. Finally, the **pdp** package (Greenwell, 2017) provides a general framework for constructing partial dependence (i.e., marginal effect) plots from various types of machine learning models in R.

Currently, besides **adabag** there are several R packages dealing with boosting methods. Among them, the most widely used are **ada** (Culp et al., 2006), **gbm** (Ridgeway, 2017), and **mboost** (Hothorn et al., 2017). These libraries are very well documented and useful. However, it seems that they are more focused on boosting for regression and dichotomous classification. The package **xgboost** (Chen and Guestrin, 2016; Chen et al., 2017), titled Extreme Gradient Boosting and maintained by Tong He, has the following description in CRAN "Extreme Gradient Boosting, which is an efficient implementation of gradient boosting framework. This package is its R interface. The package includes efficient linear model solver and tree learning algorithms. The package can automatically do parallel computation on a single machine which could be more than 10 times faster than existing gradient boosting packages. It supports various objective functions, including regression, classification and ranking. The package is made to be extensible, so that users are also allowed to define their own objectives easily". For a more detailed revision of boosting packages see the intriguing CRAN Task View: Machine Learning & Statistical Learning maintained by Torsten Hothorn (http://cran.r-project.org/web/views/MachineLearning.html).

4.2 adabag: An R Package for Classification with Boosting and Bagging

To see the advantages of the combined classifiers, we first construct an individual tree and then generate combinations of trees. The well-known iris data set from the UCI repository of machine learning databases (Dua and Karra Taniskidou, 2017) is used as an example to illustrate the use of **adabag**. This data set is the most popular one in the UCI repository with 1352281 web hits since 2007. It has 150 instances, with four numeric attributes (sepal.length, sepal.width, petal.length, petal.width) and the three classes (species of iris plant: setosa, versicolor and virginica). Although the small size of this set, 150 observations, would not advise its division in training and testing, we use it for didactic purposes. Thus, the training set will consist of 99 observations (2/3) and the other 51 will constitute the test set (1/3). To maintain a structure proportional to that of the original set, 33 observations of each class were randomly selected in the training set.

We can force the growth of a tree to achieve a perfect fit to the training set. The individual tree has the structure shown below. At each node, the following information is given, for instance, the meaning of "`2) Petal.Length< 2.6 33 0 setosa (1.00000000 0.00000000 0.00000000) *`". Firstly, the number of the split (`2`) and the condition to fall in this branch (`Petal.Length< 2.6`), only those observations with a petal length lower than 2.6 run in this node. The number `33` indicates that there are that number of observations that fulfill the mentioned condition. The next number indicates that there is no observation (`0`) in the node that is not of the majority class in it (`setosa`). Finally, the odds assigned to each class are shown in parentheses according to the proportion of observations of each of them in that node.

```
> library(rpart)
> data("iris")
> formula<-Species ~.
> vardep <- iris[,as.character(formula[[2]])]
> cntrl <- rpart.control(maxdepth = 30, minsplit = 1,  cp = -1)
> train <- c(sample(1:50, 33), sample(51:100, 33), sample(101:150, 33))
> iris.rpart<-rpart(formula=formula,method="class",data=iris[train,], control=cntrl)
> iris.rpart

n= 99

node), split, n, loss, yval, (yprob)
      * denotes terminal node

 1) root 99 66 setosa (0.33333333 0.33333333 0.33333333)
   2) Petal.Length< 2.6 33 0 setosa (1.00000000 0.00000000 0.00000000) *
   3) Petal.Length>=2.6 66 33 versicolor (0.00000000 0.50000000 0.50000000)
     6) Petal.Length< 4.75 30 1 versicolor (0.00000000 0.96666667 0.03333333)
      12) Sepal.Length>=4.95 29 0 versicolor (0.00000000 1.00000000 0.00000000) *
      13) Sepal.Length< 4.95 1 0 virginica (0.00000000 0.00000000 1.00000000) *
     7) Petal.Length>=4.75 36 4 virginica (0.00000000 0.11111111 0.88888889)
      14) Petal.Width< 1.75 6 3 versicolor (0.00000000 0.50000000 0.50000000)
        28) Petal.Length< 5.05 2 0 versicolor (0.00000000 1.00000000 0.00000000) *
        29) Petal.Length>=5.05 4 1 virginica (0.00000000 0.25000000 0.75000000)
          58) Sepal.Length< 6.05 1 0 versicolor (0.00000000 1.00000000 0.00000000) *
          59) Sepal.Length>=6.05 3 0 virginica (0.00000000 0.00000000 1.00000000) *
      15) Petal.Width>=1.75 30 1 virginica (0.00000000 0.03333333 0.96666667)
        30) Petal.Length< 4.85 3 1 virginica (0.00000000 0.33333333 0.66666667)
          60) Sepal.Length< 5.95 1 0 versicolor (0.00000000 1.00000000 0.00000000) *
          61) Sepal.Length>=5.95 2 0 virginica (0.00000000 0.00000000 1.00000000) *
        31) Petal.Length>=4.85 27 0 virginica (0.00000000 0.00000000 1.00000000) *
```

Below are the confusion matrices and the error in the training and test sets, respectively. As mentioned before this tree achieves a perfect fit to the training set, 0% training error, and a 3.92% test error.

```
> iris.predrpart<-predict(iris.rpart, newdata = iris[train,], type="class")
> table(iris.predrpart, vardep[train], dnn=c("Predicted Class", "Observed Class"))

               Observed Class
Predicted Class setosa versicolor virginica
     setosa         33          0         0
     versicolor      0         33         0
     virginica       0          0        33
```

```
> 1-sum(iris.predrpart==vardep[train])/length(vardep[train])
[1] 0

> # Test set
> iris.predrpart<-predict(iris.rpart, newdata = iris[-train,], type="class")
> table(iris.predrpart, vardep[-train], dnn=c("Predicted Class", "Observed Class"))

               Observed Class
Predicted Class setosa versicolor virginica
     setosa         17          0         0
     versicolor      0         16         1
     virginica       0          1        16

> 1-sum(iris.predrpart==vardep[-train])/length(vardep[-train])
[1] 0.03921569
```

The following table collects the output of the printcp function for this tree. First, it shows the call that has been made to the function and then the variables that have actually been used to develop the tree. In this case, the variable Sepal.Width has been left out. Subsequently, it points out the error that exists in the root node, that is, the error of the default rule (in this case 66.67%). Finally, the `cptable` shows for each cut the cost-complexity parameter, the position it occupies, the error in the training set or relative error, the cross-validation error, and the standard deviation of the cross-validation error. Both the error of the training set and the cross-validation are obtained by multiplying the value that appears in that table by the error in the root node.

```
> printcp(iris.rpart)

Classification tree:
rpart(formula = formula, data = iris[train,], method = "class",
    control = cntrl)

Variables actually used in tree construction:
[1] Petal.Length Petal.Width  Sepal.Length

Root node error: 66/99 = 0.66667

n= 99

         CP nsplit rel error   xerror     xstd
1  0.5000000      0  1.000000 1.166667 0.062675
2  0.4242424      1  0.500000 0.636364 0.074508
3  0.0151515      2  0.075758 0.106061 0.038644
4  0.0075758      6  0.015152 0.090909 0.035971
5 -1.0000000      8  0.000000 0.090909 0.035971
```

The simplicity of this example advises limiting the growth of the tree in training or pruning it once it has been developed. The reason for this is to avoid overfitting and achieve a balance between training and test errors. A common way to prune a tree is to look in the cptable for the minimum cross-validation error plus a standard deviation. The cross-validation estimation of the error (xerror) has a random component. To avoid this randomness the 1-SE rule (or 1-SD rule) selects the simplest model with a xerror equal to or less than the minimum xerror plus the standard deviation of the minimum xerror. This can be seen in Figure 4.1, which shows the plotcp of the individual tree for the iris example.

```
> plotcp(iris.rpart)
```

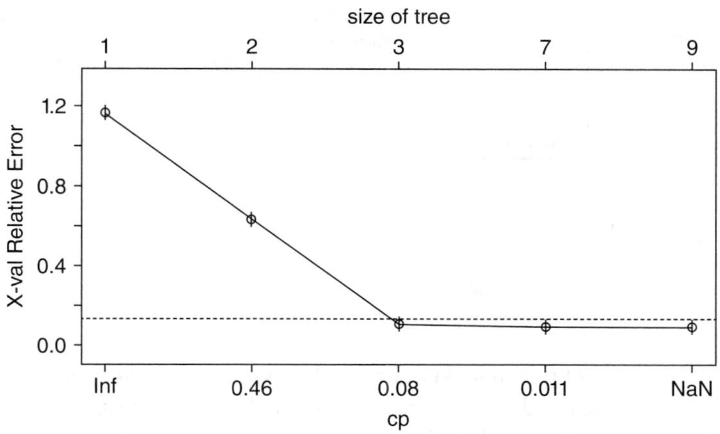

Figure 4.1 Cross-validation error versus tree complexity for the iris example.

The `autoprune` function, in the **adabag** library, automatically builds a pruned tree of class `rpart` applying the above-mentioned 1-SE rule on the xerror. The output returns an object of class `rpart`. The usage of this function is very simple, with only three arguments:

1. `formula`: a formula as in the `lm` function.
2. `data`: a data frame in which to interpret the variables named in the formula.
3. `subset`: optional expression saying that only a subset of the rows of the data should be used in the fit, as in the rpart function.
4. "...": further arguments passed to or from other methods.

Applying the `autoprune` function to the iris example, we obtain a really simple tree, which can be seen graphically in Figure 4.2.

```
> iris.prune<-autoprune(formula=formula, data=iris[train,])
> iris.prune

n= 99

node), split, n, loss, yval, (yprob)
      * denotes terminal node

1) root 99 66 setosa (0.33333333 0.33333333 0.33333333)
  2) Petal.Length< 2.6 33 0 setosa (1.00000000 0.00000000 0.00000000) *
  3) Petal.Length>=2.6 66 33 versicolor (0.00000000 0.50000000 0.50000000)
    6) Petal.Length< 4.75 30 1 versicolor (0.00000000 0.96666667 0.03333333) *
    7) Petal.Length>=4.75 36 4 virginica (0.00000000 0.11111111 0.88888889) *

> plot(iris.prune, uniform=T, margin=0.15)
> text(iris.prune, cex=0.75,use.n=TRUE)
> title(main="Individual tree structure", cex.main = 1.5, font.main= 4)
```

In this case the tree uses only the variable Petal.Length. The interpretation of this tree is that the flowers with Petal.Length< 2.6 are classified as setosa. On the other hand, if Petal.Length>= 2.6 but < 4.75, the observations are assigned to versicolor. Finally, if Petal.Length>= 4.75, the flowers are classified as virginica.

The following are the confusion matrices and the errors in the training and test sets. In this case, the error is higher in the training set (5.05%) than in the test set (3.92%). In the

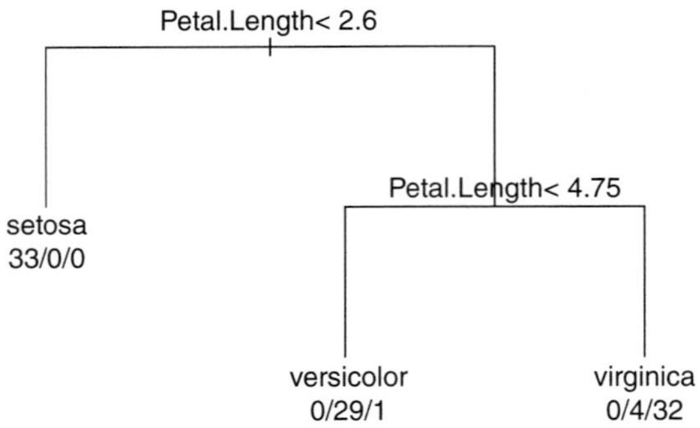

Figure 4.2 Individual tree for the iris example.

confusion matrix for the training set five mistakes can be observed, four flowers of class versicolor are classified as virginica and one virginica as versicolor. In the test set, two flowers of class versicolor are classified as virginica. As the test error of the pruned tree matches the test error of the full tree and following the Ockham's razor principle, "when you have two competing theories or models that make exactly the same predictions, the simpler one is the better", the pruned tree will be selected.

```
> # Training set
> iris.predrpart<-predict(iris.prune, newdata = iris[train,], type="class")
> table(iris.predrpart, vardep[train], dnn=c("Predicted Class", "Observed Class"))

               Observed Class
Predicted Class setosa versicolor virginica
     setosa         33          0         0
     versicolor      0         29         1
     virginica       0          4        32

> # Test set
> iris.predrpart<-predict(iris.prune, newdata = iris[-train,], type="class")
> table(iris.predrpart, vardep[-train], dnn=c("Predicted Class", "Observed Class"))

> 1-sum(iris.predrpart==vardep[train])/length(vardep[train])
[1] 0.05050505

               Observed Class
Predicted Class setosa versicolor virginica
     setosa         17          0         0
     versicolor      0         15         0
     virginica       0          2        17

> 1-sum(iris.predrpart==vardep[-train])/length(vardep[-train])
[1] 0.03921569
```

4.2.1 The `bagging`, `predict.bagging`, and `bagging.cv` Functions

Once the individual classification tree is constructed, the task of improving its results is addressed by combining trees of the same size or even smaller. To apply the algorithms bagging, AdaBoost.M1, and SAMME the library **adabag** version 4.2 (Alfaro et al., 2013; Alfaro et al., 2018) has been used. This library consists of three functions for each method, one that builds the bagging (or boosting) classifier and classifies the observations of the training set, another which uses the classifier trained to predict the class of a set of new observations, and lastly a function that allows cross-validation on a set of observations.

We start with the `bagging` function that implements the bagging algorithm. Any function in R requires a set of initial arguments that, in this case, are the following:

- `formula`: where the dependent variable and the independent variables are specified.
- `data`: the name of the data frame where the variables named in `formula` should be read. It collects the data to be used for training the ensemble.
- `mfinal`: the number of iterations or, in other words, the number of trees to use in the combination. The default value is 100.
- `control`: finally, three parameters (`maxdepth, minsplit, cp`) that are characteristic of the `rpart` function but that are also transferred to bagging to limit the size of the trees.[1]
- `par`: a logical argument, if `par=TRUE`, the cross-validation process runs in parallel. On the other hand, if `FALSE` (by default), the function is executed without parallelization.
- "...": the three dots deal with further arguments passed to or from other methods.

After applying bagging and building the combination, this function returns a bagging class object that is, in the terminology of R, a list that contains the following aspects:

- the `formula` used
- all `trees` constructed during iterations
- the sum of the `votes` that the observations receive for each class
- the `prob` matrix shows, for each observation, an approximation to the "a posteriori" probability or degree of support of each class. This estimate is calculated from the proportion of votes in the final combination.
- the `class` assigned to each observation
- the relative `importance` or contribution of each variable in the classification. This measure takes into account the gain of the Gini index given by a variable in a tree and, in the case of the boosting function, the weight of this tree. For this goal, the varImp function of the caret package (Kuhn, 2008, 2017) is used to get the gain of the Gini index of the variables in each tree.

The attributes `terms, call` and `vardep.summary` collect some information about the model and the variables used in it. The ensemble methods use a large number of trees (hundreds or maybe thousands). This hinders the interpretation of the final classifier and may cause this method to be considered as a black box. To avoid this, any tool that helps to understand the operation of the bagging or boosting classifier would be very useful. Therefore, the `bagging` and `boosting` functions have been programmed to provide a measure of the importance of the characteristics. The relative importance of the variables has been calculated from the information gain (Gini index) of each variable in the different trees and, in the case of boosting, their weights in the final combination. In order to obtain the gain in the Gini index of the variables in each tree, the function varImp of the **`caret`** package (Kuhn, 2017) is used.

Continuing with the iris example, we constructed a bagging classifier with 10 trees of maximum depth 1 using the code shown below for the training set. The output is a list with the formula used, the 10 decision stumps and the matrix votes and prob. These matrices show the votes (probabilities) that each observation (rows) receives for

1 See `rpart.control` for more details.

each class (columns). For example, the first observation receives 1 vote for the first class, 4 for the second and 5 for the third, so its posterior probabilities are 10, 40 and 50%, respectively, for each class and the assigned class is "virginica", as shown by the vector class.

```
> # bagging
> iris.bagging <- bagging(formula = Species ~., data = iris[train,], mfinal = 10,
    control = rpart.control(maxdepth = 1))

$formula
Species ~.

$trees
$trees[[1]]
n= 99

node), split, n, loss, yval, (yprob)
      * denotes terminal node

1) root 99 62 virginica (0.3232323 0.3030303 0.3737374)
  2) Petal.Length< 2.6 32  0 setosa (1.0000000 0.0000000 0.0000000) *
  3) Petal.Length>=2.6 67 30 virginica (0.0000000 0.4477612 0.5522388) *

$trees[[2]]
n= 99

node), split, n, loss, yval, (yprob)
      * denotes terminal node

1) root 99 63 virginica (0.3232323 0.3131313 0.3636364)
  2) Petal.Length< 2.6 32  0 setosa (1.0000000 0.0000000 0.0000000) *
  3) Petal.Length>=2.6 67 31 virginica (0.0000000 0.4626866 0.5373134) *

$trees[[3]]
n= 99

node), split, n, loss, yval, (yprob)
      * denotes terminal node

1) root 99 62 virginica (0.3131313 0.3131313 0.3737374)
  2) Petal.Length< 2.6 31  0 setosa (1.0000000 0.0000000 0.0000000) *
  3) Petal.Length>=2.6 68 31 virginica (0.0000000 0.4558824 0.5441176) *

$trees[[4]]
n= 99

node), split, n, loss, yval, (yprob)
      * denotes terminal node

1) root 99 61 versicolor (0.3131313 0.3838384 0.3030303)
  2) Petal.Length< 2.6 31  0 setosa (1.0000000 0.0000000 0.0000000) *
  3) Petal.Length>=2.6 68 30 versicolor (0.0000000 0.5588235 0.4411765) *

$trees[[5]]
n= 99

node), split, n, loss, yval, (yprob)
      * denotes terminal node

1) root 99 60 versicolor (0.21212121 0.39393939 0.39393939)
  2) Petal.Width< 1.75 63 25 versicolor (0.33333333 0.60317460 0.06349206) *
  3) Petal.Width>=1.75 36  1 virginica (0.00000000 0.02777778 0.97222222) *
```

```
$trees[[6]]
n= 99

node), split, n, loss, yval, (yprob)
      * denotes terminal node

1) root 99 58 versicolor (0.3232323 0.4141414 0.2626263)
  2) Petal.Length< 2.7 32  0 setosa (1.0000000 0.0000000 0.0000000) *
  3) Petal.Length>=2.7 67 26 versicolor (0.0000000 0.6119403 0.3880597) *

$trees[[7]]
n= 99

node), split, n, loss, yval, (yprob)
      * denotes terminal node

1) root 99 61 versicolor (0.29292929 0.38383838 0.32323232)
  2) Petal.Width< 1.75 68 30 versicolor (0.42647059 0.55882353 0.01470588) *
  3) Petal.Width>=1.75 31  0 virginica (0.00000000 0.00000000 1.00000000) *

$trees[[8]]
n= 99

node), split, n, loss, yval, (yprob)
      * denotes terminal node

1) root 99 64 versicolor (0.3333333 0.3535354 0.3131313)
  2) Petal.Length< 2.5 33  0 setosa (1.0000000 0.0000000 0.0000000) *
  3) Petal.Length>=2.5 66 31 versicolor (0.0000000 0.5303030 0.4696970) *

$trees[[9]]
n= 99

node), split, n, loss, yval, (yprob)
      * denotes terminal node

1) root 99 59 versicolor (0.24242424 0.40404040 0.35353535)
  2) Petal.Width< 1.75 68 28 versicolor (0.35294118 0.58823529 0.05882353) *
  3) Petal.Width>=1.75 31  0 virginica (0.00000000 0.00000000 1.00000000) *

$trees[[10]]
n= 99

node), split, n, loss, yval, (yprob)
      * denotes terminal node

1) root 99 61 virginica (0.2929293 0.3232323 0.3838384)
  2) Petal.Length< 4.75 59 29 versicolor (0.4915254 0.5084746 0.0000000) *
  3) Petal.Length>=4.75 40  2 virginica (0.0000000 0.0500000 0.9500000) *

$votes
       [,1] [,2] [,3]
  [1,]    6    4    0
  [2,]    6    4    0
  [3,]    6    4    0
 ...
 [33,]    6    4    0
 [34,]    0    7    3
 [35,]    0    7    3
 ...
 [66,]    0    7    3
 [67,]    0    3    7
```

```
 [68,]    0    3    7
 ...
 [97,]    0    3    7
 [98,]    0    6    4
 [99,]    0    3    7

$prob
      [,1] [,2] [,3]
  [1,]  0.6  0.4  0.0
  [2,]  0.6  0.4  0.0
  [3,]  0.6  0.4  0.0
...
 [33,]  0.6  0.4  0.0
 [34,]  0.0  0.7  0.3
 [35,]  0.0  0.7  0.3
...
 [66,]  0.0  0.7  0.3
 [67,]  0.0  0.3  0.7
 [68,]  0.0  0.3  0.7
...
 [97,]  0.0  0.3  0.7
 [98,]  0.0  0.6  0.4
 [99,]  0.0  0.3  0.7

$class
[1] "setosa"     "setosa"     "setosa"     "setosa"     "setosa"
 [6] "setosa"     "setosa"     "setosa"     "setosa"     "setosa"
[11] "setosa"     "setosa"     "setosa"     "setosa"     "setosa"
[16] "setosa"     "setosa"     "setosa"     "setosa"     "setosa"
[21] "setosa"     "setosa"     "setosa"     "setosa"     "setosa"
[26] "setosa"     "setosa"     "setosa"     "setosa"     "setosa"
[31] "setosa"     "setosa"     "setosa"     "versicolor" "versicolor"
[36] "versicolor" "versicolor" "virginica"  "versicolor" "versicolor"
[41] "versicolor" "versicolor" "versicolor" "versicolor" "versicolor"
[46] "versicolor" "versicolor" "versicolor" "versicolor" "versicolor"
[51] "versicolor" "versicolor" "versicolor" "versicolor" "versicolor"
[56] "versicolor" "versicolor" "versicolor" "versicolor" "versicolor"
[61] "versicolor" "versicolor" "versicolor" "versicolor" "versicolor"
[66] "versicolor" "virginica"  "virginica"  "virginica"  "virginica"
[71] "virginica"  "virginica"  "versicolor" "virginica"  "virginica"
[76] "virginica"  "virginica"  "virginica"  "virginica"  "virginica"
[81] "virginica"  "virginica"  "virginica"  "virginica"  "virginica"
[86] "virginica"  "virginica"  "versicolor" "virginica"  "virginica"
[91] "virginica"  "virginica"  "versicolor" "virginica"  "virginica"
[96] "virginica"  "virginica"  "versicolor" "virginica"

$samples
      [,1] [,2] [,3] [,4] [,5] [,6] [,7] [,8] [,9] [,10]
  [1,]   49   61   75   23   58   43   45   50   84    91
  [2,]   49   29   82   65   48   38   61   35   83    65
  [3,]   52    9   66   66   45   36    3   28   30    41
...
 [33,]   94   99   56   20   74   98   22   53   56    42
 [34,]   48   35   68   73   34   41   19   29   48    87
 [35,]   83   95    7   28    7   84   83   53   89    86
...
 [66,]   73   47    5   70   80   21    8   64   92    23
 [67,]   56   77   20   85   46   50   60   32   95    89
 [68,]   96   36   79   39   16   50   26   99   83    99
...
 [97,]   37   54   15   28   17   48   39   31   61    28
```

```
 [98,]    76    47    65    18    56    59    85    80    50     14
 [99,]    38    52    95    89    80    97    67    29    23     37

$importance
Petal.Length  Petal.Width Sepal.Length  Sepal.Width
    71.86163     28.13837      0.00000      0.00000

$terms
Species ~ Sepal.Length + Sepal.Width + Petal.Length + Petal.Width
attr(,"variables")
list(Species, Sepal.Length, Sepal.Width, Petal.Length, Petal.Width)
attr(,"factors")
             Sepal.Length Sepal.Width Petal.Length Petal.Width
Species                 0           0            0           0
Sepal.Length            1           0            0           0
Sepal.Width             0           1            0           0
Petal.Length            0           0            1           0
Petal.Width             0           0            0           1
attr(,"term.labels")
[1] "Sepal.Length" "Sepal.Width"  "Petal.Length" "Petal.Width"
attr(,"order")
[1] 1 1 1 1
attr(,"intercept")
[1] 1
attr(,"response")
[1] 1
attr(,".Environment")
<environment: R_GlobalEnv>
attr(,"predvars")
list(Species, Sepal.Length, Sepal.Width, Petal.Length, Petal.Width)
attr(,"dataClasses")
     Species Sepal.Length  Sepal.Width Petal.Length  Petal.Width
    "factor"    "numeric"    "numeric"    "numeric"    "numeric"

$call
bagging(formula = Species ~., data = iris[train,], mfinal = 10,
    control = rpart.control(maxdepth = 1))

attr(,"vardep.summary")
    setosa versicolor  virginica
        33         33         33
attr(,"class")
[1] "bagging"
```

The sample matrix collects the boostrap replicates used in each iteration. It can be seen how observations 24 and 39 are repeated four times in the first of these replicates and some observations appear two or three times. On the contrary, observations 5 and 7, among others, are left out and, therefore, do not participate in the training of the first tree. The variable Petal.Length is the one that contributes the most, obtaining a 71.86% of the information gain, followed again by the variable Petal.Width (28.14%). On the contrary, the other two variables do not contribute anything. This can be seen graphically in Figure 4.3.

```
> iris.bagging$samples[,1]

 [1] 49 49 52  8 97 32 48 19  1 26 24  6 84 49 97  9 48 99  4 50 43 39 39 35 71 75 43  8
[29] 54 54 62 93 94 48 83 27 72 28 24 25 18 65 30 89 95 33 68 72 76  3 84  9 46 33 93 89
[57] 99 94 21 51 39 98 98 24 73 73 56 96 50 15 47 88 62 26 92  8 82 80 31 68 85 24 39 23
[85] 19 44 23 77 98  2 20 60 97 92 83 43 37 76 38
```

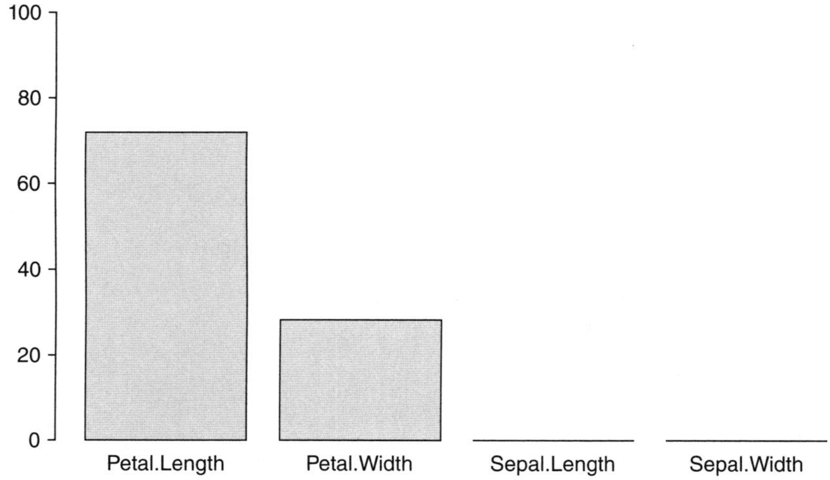

Figure 4.3 Variable relative importance in bagging for the iris example.

```
> summary(as.factor(iris.bagging$samples[,1]))

 1  2  3  4  6  8  9 15 18 19 20 21 23 24 25 26 27 28 30 31 32 33 35 37 38 39 43 44 46
 1  1  1  1  1  3  2  1  1  2  1  1  2  4  1  2  1  1  1  1  1  2  1  1  1  4  3  1  1
47 48 49 50 51 52 54 56 60 62 65 68 71 72 73 75 76 77 80 82 83 84 85 88 89 92 93 94 95
 1  3  3  2  1  1  2  1  1  2  1  2  1  2  2  1  2  1  1  1  2  2  1  1  2  2  2  2  1
96 97 98 99
 1  3  3  2
```

Since version 4.0 **adabag** has included three new plot functions. One of them, `importanceplot()`, draws the relative importance of each variable in the classification task. The main argument in this function is `object`, a fitted model object of class bagging or boosting. This is assumed to be the result of some function that produces an object with a component named `importance` as that returned by those functions. On the other hand, ...indicates further arguments passed to or from other methods. A labeled plot is produced on the current graphics device (one being opened if needed).

```
> importanceplot(iris.bagging, horiz=F, main=" Variable Relative Importance")
```

Using the object `iris.bagging` is easy to build the confusion matrix for the training set and calculate its error. In this case, bagging fails in five examples, one flower of class versicolor is classified as virginica and four virginica as versicolor. Therefore, the training error is 5.05%.

```
> # Training set
> table(iris.bagging$class, iris$Species[train], dnn=c("Predicted Class",
  "Observed Class"))

               Observed Class
Predicted Class setosa versicolor virginica
     setosa         33          0         0
     versicolor      0         32         4
     virginica       0          1        29
> 1-sum(data.bagging$class==data$clase[ind])/length(data$clase[ind])
[1] 0.05050505
```

The second function, `predict.bagging`, follows the parameters of the generic `predict` function but adds one more (`object, newdata, newmfinal=length (object$trees), ...`). This function is used to classify a data frame using an adjusted model in a bagging class object. Its input parameters are:

- `object`: the result of some function that produces an object whose components have the same name as those returned by the `bagging` function.
- `newdata`: a data frame with the values that should be classified. The variables mentioned in the right part of `formula (object)` must be here with the same name. In this way, predictions can be made beyond the training set.
- `newmfinal`: this option sets the number of `bagging` object trees to be used in the prediction. This allows the user to prune the combination. By default, all object trees are used.
- "...": These deal with further arguments passed to or from other methods.

In contrast, this function returns a `predict.bagging` class object, which is a list of six components. Of these, the first four (`formula, votes, prob` and `class`) have the same meaning as in `bagging` output. The other two are:

- `confusion`: the confusion matrix that compares the real class with the prediction or NULL when predicting with unlabeled data.
- `error`: the error rate or NULL when predicting with unlabeled data.

In the iris example, the `predict.bagging` function uses the previously trained `iris.bagging` object to predict the class of the test set flowers and, if necessary, prune it. In this case, the confusion matrix shows only an error, a virginica flower classified as versicolor. Therefore, the test error is 1.96%.

```
> #Test set
>   iris.predbagging<- predict.bagging(iris.bagging, newdata=iris[-train,])
>   iris.predbagging

$formula
Species ~.

$votes
       [,1] [,2] [,3]
 [1,]     6    4    0
 [2,]     6    4    0
 [3,]     6    4    0
 ...
[17,]     6    4    0
[18,]     0    7    3
[19,]     0    7    3
...
[34,]     0    7    3
[35,]     0    3    7
[36,]     0    3    7
...
[49,]     0    3    7
[50,]     0    3    7
[51,]     0    3    7

$prob
```

```
      [,1] [,2] [,3]
 [1,]  0.6  0.4  0.0
 [2,]  0.6  0.4  0.0
 [3,]  0.6  0.4  0.0
...
[17,]  0.6  0.4  0.0
[18,]  0.0  0.7  0.3
[19,]  0.0  0.7  0.3
...
[34,]  0.0  0.7  0.3
[35,]  0.0  0.3  0.7
[36,]  0.0  0.3  0.7
...
[49,]  0.0  0.3  0.7
[50,]  0.0  0.3  0.7
[51,]  0.0  0.3  0.7

$class
 [1] "setosa"     "setosa"     "setosa"     "setosa"     "setosa"     "setosa"
 [7] "setosa"     "setosa"     "setosa"     "setosa"     "setosa"     "setosa"
[13] "setosa"     "setosa"     "setosa"     "setosa"     "setosa"     "versicolor"
[19] "versicolor" "versicolor" "versicolor" "versicolor" "versicolor" "versicolor"
[25] "versicolor" "versicolor" "versicolor" "versicolor" "versicolor" "versicolor"
[31] "versicolor" "versicolor" "versicolor" "versicolor" "virginica"  "virginica"
[37] "virginica"  "virginica"  "virginica"  "versicolor" "virginica"  "virginica"
[43] "virginica"  "virginica"  "virginica"  "virginica"  "virginica"  "virginica"
[49] "virginica"  "virginica"  "virginica"

$confusion
               Observed Class
Predicted Class setosa versicolor virginica
     setosa         17          0         0
     versicolor      0         17         1
     virginica       0          0        16

$error
[1] 0.01960784
```

Finally, the third function `bagging.cv` runs v-fold cross-validation with bagging. As usually in cross-validation, the data are divided into v non-overlapping subsets of roughly equal size. Then, bagging is applied on $(v - 1)$ of the subsets. Lastly, predictions are made for the left-out subset, and the process is repeated for each one of the v subsets.

Five of the arguments of this function are the same as bagging and one more, v. The integer v specifies the type of v-fold cross-validation. The default value is 10. In the extreme case that v is equal to the number of observations, leave-one-out cross-validation is carried out. Besides this, every value between two and the number of observations is valid and means that one out of every v observations is left out. Therefore, we can use cross-validation to estimate the error of a classifier without dividing the data available in training and the test. This is especially interesting for small data sets.

An object of class `bagging.cv` is supplied, which is a list with three components, `class`, `confusion` and `error`, which have been previously described in the `predict.bagging` function.

Next, we estimate the bagging error by cross-validation in 10 folders, maintaining the number and size of the trees. On this occasion, 23 virginica flowers are erroneously classified as versicolor. The model was also wrong to classify six versicolor flowers as virginica. Therefore, the error estimate we obtain is 19.33%.

```
> iris.baggingcv <- bagging.cv(Species ~., v=10, data=iris, mfinal=10,
  control=rpart.control(maxdepth=1))
>   iris.baggingcv

$class
  [1] "setosa"     "setosa"     "setosa"     "setosa"     "setosa"     "setosa"
  [7] "setosa"     "setosa"     "setosa"     "setosa"     "setosa"     "setosa"
 [13] "setosa"     "setosa"     "setosa"     "setosa"     "setosa"     "setosa"
 [19] "setosa"     "setosa"     "setosa"     "setosa"     "setosa"     "setosa"
 [25] "setosa"     "setosa"     "setosa"     "setosa"     "setosa"     "setosa"
 [31] "setosa"     "setosa"     "setosa"     "setosa"     "setosa"     "setosa"
 [37] "setosa"     "setosa"     "setosa"     "setosa"     "setosa"     "setosa"
 [43] "setosa"     "setosa"     "setosa"     "setosa"     "setosa"     "setosa"
 [49] "setosa"     "setosa"     "versicolor" "virginica"  "versicolor" "versicolor"
 [55] "versicolor" "versicolor" "versicolor" "versicolor" "versicolor" "versicolor"
 [61] "versicolor" "virginica"  "versicolor" "versicolor" "versicolor" "versicolor"
 [67] "versicolor" "versicolor" "versicolor" "versicolor" "virginica"  "virginica"
 [73] "versicolor" "versicolor" "versicolor" "versicolor" "versicolor" "versicolor"
 [79] "versicolor" "versicolor" "versicolor" "virginica"  "versicolor" "versicolor"
 [85] "versicolor" "versicolor" "versicolor" "versicolor" "versicolor" "versicolor"
 [91] "versicolor" "virginica"  "versicolor" "versicolor" "versicolor" "versicolor"
 [97] "versicolor" "versicolor" "versicolor" "versicolor" "virginica"  "virginica"
[103] "versicolor" "versicolor" "virginica"  "virginica"  "versicolor" "versicolor"
[109] "virginica"  "virginica"  "virginica"  "virginica"  "versicolor" "versicolor"
[115] "virginica"  "virginica"  "versicolor" "versicolor" "virginica"  "versicolor"
[121] "virginica"  "virginica"  "versicolor" "versicolor" "virginica"  "virginica"
[127] "versicolor" "versicolor" "virginica"  "versicolor" "virginica"  "virginica"
[133] "versicolor" "versicolor" "versicolor" "virginica"  "versicolor" "versicolor"
[139] "virginica"  "virginica"  "virginica"  "virginica"  "versicolor" "versicolor"
[145] "virginica"  "virginica"  "versicolor" "versicolor" "virginica"  "virginica"

$confusion
               Observed Class
Predicted Class setosa versicolor virginica
     setosa         50          0         0
     versicolor      0         44        23
     virginica       0          6        27

$error
[1] 0.1933333
```

4.2.2 The `boosting`, `predict.boosting`, and `boosting.cv` Functions

Similarly to bagging, boosting also has three functions related to it in the **adabag** package. The `boosting` function implements the AdaBoost.M1 and SAMME algorithms. It constructs this type of combined classifier and assigns the class to the observations of the training set. Five of the initial parameters (`formula, data, mfinal, control, ...`) have the same meaning as in bagging, whereas two arguments, `boos` and `coeflearn`, are new. It should be highlighted that, unlike in bagging, the individual classifiers are now not statistically independent.

- `boos`: a logical indicator that shows whether you want to use the resampling technique (TRUE, by default) or reweighting (FALSE).
- `coeflearn`: a logical argument, if 'Breiman' is chosen (by default), then $\alpha = \frac{1}{2}\ln\left(\frac{1-e_b}{e_b}\right)$ is used. If `coeflearn` = 'Freund', $\alpha = \ln\left(\frac{1-e_b}{e_b}\right)$. In both cases

the algorithm AdaBoost.M1 is used and alpha is the update constant of the weights. On the other hand, if "Zhu" is set, the SAMME algorithm is implemented with $\alpha = \ln\left(\frac{1-e_b}{e_b}\right) + \ln(k-1)$. As mentioned above, the error of the individual trees must be between (0, 0.5) for AdaBoost.M1 and between $(0, 1 - 1/k)$ for SAMME. In the event that this is breached, Opitz and Maclin (1999) equalize all weights, choose an appropriate value to weight this tree, and restart the process. The same solution applies here. When $e_b = 0$, the alpha constant is calculated using $e_b = 0.001$ and when $e_b \geq 0.5$ ($e_b \geq 1 - 1/k$ for SAMME), it is changed to 0.499 $(0.999 - 1/k)$.

This function produces an object of class `boosting`, which is a list with almost the same components as `bagging`. The only difference is the matrix with the weights of the trees instead of the bootstrap samples used along the iterations.

Continuing with the iris example, we classify the training set with a boosting combination of 10 trees of maximum depth 1, or decision stumps, using the code shown below. The output is a list with the formula used, the 10 small trees and their weights. The smaller the error of a tree, the greater its weight, that is, the more important it is in the final combination. In this case, the maximum weight is for the fourth tree (0.557) and the smallest is the third tree (0.193). The votes and prob matrices show the sum of the weighted votes and the probabilities, respectively, that each observation (rows) receives for each class (columns). For example, the first observation receives 2.247 votes for the first, 1.193 for the second and 0 for the third class, so its a posteriori odds are 65.315, 34.685 and 0%, respectively, for each class and the assigned class is setosa, as shown by the class vector.

```
> # boosting
> library(adabag)
> iris.adaboost <- boosting(Species ~., data=iris[train,], mfinal=10,
+ control=rpart.control(maxdepth=1))
> iris.adaboost

$formula
Species ~.
$trees
$trees[[1]]
n= 99
node), split, n, loss, yval, (yprob)
      * denotes terminal node
1) root 99 64 versicolor (0.2929293 0.3535354 0.3535354)
  2) Petal.Length< 2.5 29  0 setosa (1.0000000 0.0000000 0.0000000) *
  3) Petal.Length>=2.5 70 35 versicolor (0.0000000 0.5000000 0.5000000) *
$trees[[2]]
n= 99
node), split, n, loss, yval, (yprob)
      * denotes terminal node
1) root 99 59 virginica (0.2222222 0.3737374 0.4040404)
  2) Petal.Length< 4.95 62 25 versicolor (0.3548387 0.5967742 0.0483871) *
  3) Petal.Length>=4.95 37  0 virginica (0.0000000 0.0000000 1.0000000) *
$trees[[3]]
n= 99
node), split, n, loss, yval, (yprob)
      * denotes terminal node
1) root 99 62 virginica (0.2626263 0.3636364 0.3737374)
  2) Petal.Width< 1.75 64 28 versicolor (0.4062500 0.5625000 0.0312500) *
  3) Petal.Width>=1.75 35  0 virginica (0.0000000 0.0000000 1.0000000) *
...
$trees[[8]]
```

```
n= 99

node), split, n, loss, yval, (yprob)
      * denotes terminal node
1) root 99 59 virginica (0.24242424 0.35353535 0.40404040)
  2) Petal.Width< 1.65 59 25 versicolor (0.40677966 0.57627119 0.01694915) *
  3) Petal.Width>=1.65 40  1 virginica (0.00000000 0.02500000 0.97500000) *
$trees[[9]]
n= 99
node), split, n, loss, yval, (yprob)
      * denotes terminal node
1) root 99 61 versicolor (0.2828283 0.3838384 0.3333333)
  2) Petal.Length< 2.75 28  0 setosa (1.0000000 0.0000000 0.0000000) *
  3) Petal.Length>=2.75 71 33 versicolor (0.0000000 0.5352113 0.4647887) *
$trees[[10]]
n= 99
node), split, n, loss, yval, (yprob)
      * denotes terminal node
1) root 99 51 virginica (0.2424242 0.2727273 0.4848485)
  2) Petal.Width< 1.75 57 30 versicolor (0.4210526 0.4736842 0.1052632) *
  3) Petal.Width>=1.75 42  0 virginica (0.0000000 0.0000000 1.0000000) *

$weights
 [1] 0.3465736 0.2858340 0.1929224 0.5565577 0.4064207 0.3868016 0.3141397 0.4478705
 [9] 0.2366235 0.2666908
$votes
          [,1]     [,2]      [,3]
  [1,] 2.247117 1.193318 0.0000000
  [2,] 2.247117 1.193318 0.0000000
  [3,] 2.247117 1.193318 0.0000000
 ...
 [33,] 2.247117 1.193318 0.0000000
 [34,] 0.000000 2.497075 0.9433593
 [35,] 0.000000 2.497075 0.9433593
 ...
 [66,] 0.000000 2.497075 0.9433593
 [67,] 0.000000 1.303757 2.1366770
 [68,] 0.000000 1.303757 2.1366770
 ...
 [97,] 0.000000 1.303757 2.1366770
 [98,] 0.000000 2.211241 1.2291933
 [99,] 0.000000 1.303757 2.1366770

$prob
          [,1]      [,2]      [,3]
  [1,] 0.6531491 0.3468509 0.0000000
  [2,] 0.6531491 0.3468509 0.0000000
  [3,] 0.6531491 0.3468509 0.0000000
  ...
 [33,] 0.6531491 0.3468509 0.0000000
 [34,] 0.0000000 0.7258023 0.2741977
 [35,] 0.0000000 0.7258023 0.2741977
...
 [66,] 0.0000000 0.7258023 0.2741977
 [67,] 0.0000000 0.3789514 0.6210486
 [68,] 0.0000000 0.3789514 0.6210486
...
 [97,] 0.0000000 0.3789514 0.6210486
 [98,] 0.0000000 0.6427215 0.3572785
 [99,] 0.0000000 0.3789514 0.6210486

$class
```

```
 [1] "setosa"     "setosa"     "setosa"     "setosa"     "setosa"     "setosa"
 [7] "setosa"     "setosa"     "setosa"     "setosa"     "setosa"     "setosa"
[13] "setosa"     "setosa"     "setosa"     "setosa"     "setosa"     "setosa"
[19] "setosa"     "setosa"     "setosa"     "setosa"     "setosa"     "setosa"
[25] "setosa"     "setosa"     "setosa"     "setosa"     "setosa"     "setosa"
[31] "setosa"     "setosa"     "setosa"     "versicolor" "versicolor" "versicolor"
[37] "versicolor" "virginica"  "versicolor" "versicolor" "versicolor" "versicolor"
[43] "versicolor" "versicolor" "versicolor" "versicolor" "versicolor" "versicolor"
[49] "versicolor" "versicolor" "versicolor" "versicolor" "versicolor" "versicolor"
[55] "versicolor" "versicolor" "versicolor" "versicolor" "versicolor" "versicolor"
[61] "versicolor" "versicolor" "versicolor" "versicolor" "versicolor" "versicolor"
[67] "virginica"  "virginica"  "virginica"  "virginica"  "virginica"  "virginica"
[73] "versicolor" "virginica"  "virginica"  "virginica"  "virginica"  "virginica"
[79] "virginica"  "virginica"  "virginica"  "virginica"  "virginica"  "virginica"
[85] "virginica"  "virginica"  "virginica"  "versicolor" "virginica"  "virginica"
[91] "virginica"  "virginica"  "versicolor" "virginica"  "virginica"  "virginica"
[97] "virginica"  "versicolor" "virginica"

$importance
Petal.Length  Petal.Width Sepal.Length  Sepal.Width
     73.8452      26.1548       0.0000       0.0000
$terms
Species ~ Sepal.Length + Sepal.Width + Petal.Length + Petal.Width
attr(,"variables")
list(Species, Sepal.Length, Sepal.Width, Petal.Length, Petal.Width)
attr(,"factors")
             Sepal.Length Sepal.Width Petal.Length Petal.Width
Species                 0           0            0           0
Sepal.Length            1           0            0           0
Sepal.Width             0           1            0           0
Petal.Length            0           0            1           0
Petal.Width             0           0            0           1
attr(,"term.labels")
[1] "Sepal.Length" "Sepal.Width"  "Petal.Length" "Petal.Width"
attr(,"order")
[1] 1 1 1 1
attr(,"intercept")
[1] 1
attr(,"response")
[1] 1
attr(,".Environment")
<environment: R_GlobalEnv>
attr(,"predvars")
list(Species, Sepal.Length, Sepal.Width, Petal.Length, Petal.Width)
attr(,"dataClasses")
     Species Sepal.Length  Sepal.Width Petal.Length  Petal.Width
    "factor"    "numeric"    "numeric"    "numeric"    "numeric"

$call
boosting(formula = Species ~., data = iris[train,], mfinal = 10,
    control = rpart.control(maxdepth = 1))

attr(,"vardep.summary")
    setosa versicolor  virginica
        33         33         33
attr(,"class")
[1] "boosting"
```

As in bagging, the importance vector reports that the variables that contribute the most are Petal.Length, since it achieves 73.845% of the information gain, and

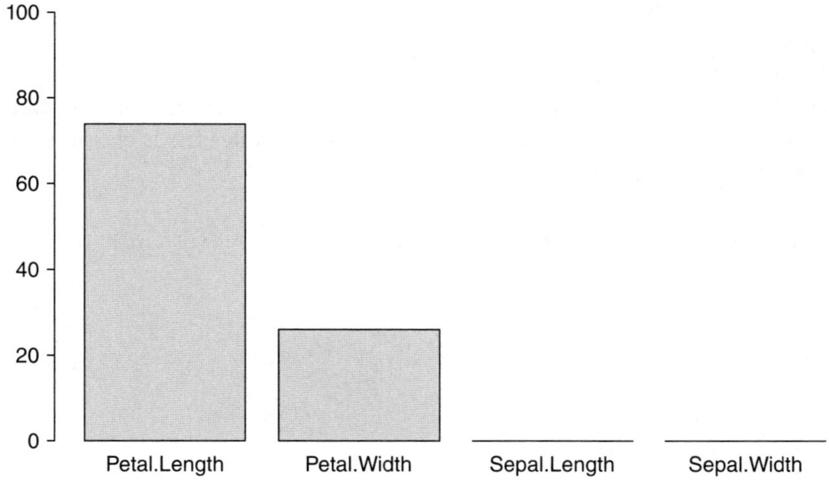

Figure 4.4 Variable relative importance in boosting for the iris example.

Petal.Width (24.34%). The two variables that measure the sepal do not contribute any of the gain in Gini index. Figure 4.4 shows each variable contribution.

```
> importanceplot(iris.adaboost, horiz=F, main=" Variable Relative Importance")
```

As previously mentioned, the confusion matrix can be constructed and the error calculated for the training set using `iris.adaboost`. In this case, boosting fails in five examples, one flower of class versicolor is classified as virginica and four virginica as versicolor. Therefore, the training error is 5.05%.

```
> #Training error
> table(iris.adaboost$class, iris$Species[train], dnn=c("Predicted Class",
+ "Observed Class"))

               Observed Class
Predicted Class setosa versicolor virginica
     setosa         33          0         0
     versicolor      0         32         4
     virginica       0          1        29

>   1-sum(iris.adaboost$class==iris$Species[train])/length(iris$Species[train])
[1] 0.05050505
```

In addition, `predict.boosting` has the same arguments and values as `predict.bagging`. However, this function calls a `boosting` class object and returns another of class `predict.boosting`. Finally, `boosting.cv` executes boosting with cross-validation in v folders. Again, the number v of chosen cross-validations is added to the arguments of the `boosting` function. In addition, the output is an object of class `boosting.cv` with the same structure as `bagging.cv`. It should be pointed out that if the logical argument `par = TRUE`, the cross-validation process runs in parallel. This option was included in version 4.2 and could be really helpful for saving time when dealing with big data.

Following the example, the `iris.adaboost` classifier can be used to predict the class for new iris examples by means of the `predict.boosting` function, as is shown below, and the ensemble can also be pruned. The first four components of the output are common to the previous function output but the confusion matrix and the test data set error are here additionally provided. In the test set, one virginica iris is classified as versicolor class, so the test error reaches 1.96%.

```
> #Test error
>   iris.predboosting<- predict.boosting(iris.adaboost, newdata=iris[-train,])
>   iris.predboosting

$formula
Species ~.

$votes
          [,1]     [,2]      [,3]
[1,] 2.247117 1.193318 0.0000000
[2,] 2.247117 1.193318 0.0000000
[3,] 2.247117 1.193318 0.0000000
...
[17,] 2.247117 1.193318 0.0000000
[18,] 0.000000 2.497075 0.9433593
[19,] 0.000000 2.497075 0.9433593
...
[34,] 0.000000 2.497075 0.9433593
[35,] 0.000000 1.303757 2.1366770
[36,] 0.000000 1.303757 2.1366770
...
[49,] 0.000000 1.303757 2.1366770
[50,] 0.000000 1.303757 2.1366770
[51,] 0.000000 1.303757 2.1366770

$prob
           [,1]      [,2]      [,3]
[1,] 0.6531491 0.3468509 0.0000000
[2,] 0.6531491 0.3468509 0.0000000
[3,] 0.6531491 0.3468509 0.0000000
 ...
[17,] 0.6531491 0.3468509 0.0000000
[18,] 0.0000000 0.7258023 0.2741977
[19,] 0.0000000 0.7258023 0.2741977
...
[34,] 0.0000000 0.7258023 0.2741977
[35,] 0.0000000 0.3789514 0.6210486
[36,] 0.0000000 0.3789514 0.6210486
...
[49,] 0.0000000 0.3789514 0.6210486
[50,] 0.0000000 0.3789514 0.6210486
[51,] 0.0000000 0.3789514 0.6210486

$class
 [1] "setosa"     "setosa"     "setosa"     "setosa"     "setosa"     "setosa"
 [7] "setosa"     "setosa"     "setosa"     "setosa"     "setosa"     "setosa"
[13] "setosa"     "setosa"     "setosa"     "setosa"     "setosa"     "versicolor"
[19] "versicolor" "versicolor" "versicolor" "versicolor" "versicolor" "versicolor"
[25] "versicolor" "versicolor" "versicolor" "versicolor" "versicolor" "versicolor"
[31] "versicolor" "versicolor" "versicolor" "versicolor" "virginica"  "virginica"
[37] "virginica"  "virginica"  "virginica"  "versicolor" "virginica"  "virginica"
[43] "virginica"  "virginica"  "virginica"  "virginica"  "virginica"  "virginica"
[49] "virginica"  "virginica"  "virginica"
```

```
$confusion
               Observed Class
Predicted Class setosa versicolor virginica
     setosa         17          0         0
     versicolor      0         17         1
     virginica       0          0        16

$error
[1] 0.01960784
```

Next, we estimate the boosting error by cross-validation in 10 folders, maintaining the number and size of the trees. On this occasion, there are four virginica flowers classified as versicolor and three versicolor classified as virginica, so the estimated error is 4.67%.

```
> iris.boostcv <- boosting.cv(Species ~., v=10, data=iris, mfinal=10,
+ control=rpart.control(maxdepth=1))
>   iris.boostcv

$class
  [1] "setosa"     "setosa"     "setosa"     "setosa"     "setosa"     "setosa"
  [7] "setosa"     "setosa"     "setosa"     "setosa"     "setosa"     "setosa"
 [13] "setosa"     "setosa"     "setosa"     "setosa"     "setosa"     "setosa"
 [19] "setosa"     "setosa"     "setosa"     "setosa"     "setosa"     "setosa"
 [25] "setosa"     "setosa"     "setosa"     "setosa"     "setosa"     "setosa"
 [31] "setosa"     "setosa"     "setosa"     "setosa"     "setosa"     "setosa"
 [37] "setosa"     "setosa"     "setosa"     "setosa"     "setosa"     "setosa"
 [43] "setosa"     "setosa"     "setosa"     "setosa"     "setosa"     "setosa"
 [49] "setosa"     "setosa"     "versicolor" "versicolor" "versicolor" "versicolor"
 [55] "versicolor" "versicolor" "versicolor" "versicolor" "versicolor" "versicolor"
 [61] "versicolor" "versicolor" "versicolor" "versicolor" "versicolor" "versicolor"
 [67] "versicolor" "versicolor" "versicolor" "versicolor" "virginica"  "versicolor"
 [73] "versicolor" "versicolor" "versicolor" "versicolor" "versicolor" "virginica"
 [79] "versicolor" "versicolor" "versicolor" "versicolor" "versicolor" "virginica"
 [85] "versicolor" "versicolor" "versicolor" "versicolor" "versicolor" "versicolor"
 [91] "versicolor" "versicolor" "versicolor" "versicolor" "versicolor" "versicolor"
 [97] "versicolor" "versicolor" "versicolor" "versicolor" "virginica"  "virginica"
[103] "virginica"  "virginica"  "virginica"  "virginica"  "versicolor" "virginica"
[109] "virginica"  "virginica"  "virginica"  "virginica"  "virginica"  "virginica"
[115] "virginica"  "virginica"  "virginica"  "virginica"  "virginica"  "versicolor"
[121] "virginica"  "virginica"  "virginica"  "versicolor" "virginica"  "virginica"
[127] "virginica"  "virginica"  "virginica"  "virginica"  "virginica"  "virginica"
[133] "virginica"  "virginica"  "versicolor" "virginica"  "virginica"  "virginica"
[139] "virginica"  "virginica"  "virginica"  "virginica"  "virginica"  "virginica"
[145] "virginica"  "virginica"  "virginica"  "virginica"  "virginica"  "virginica"

$confusion
               Observed Class
Predicted Class setosa versicolor virginica
     setosa         50          0         0
     versicolor      0         47         4
     virginica       0          3        46

$error
[1] 0.04666667
```

4.2.3 The `margins`, `plot.margins`, `errorevol` and `plot.errorevol` Functions

The margins function allows us to calculate the margin achieved by applying bagging or boosting to a given set, as defined in section 3.4.2. Its use is quite simple since it only has the following two arguments:

- `object`: the output of some of the functions bagging, boosting, `predict.bagging` or `predict.boosting`. It is the result of some function with at least two components, `formula` and `class`, as in the `bagging` function, for example.
- `newdata`: a data set with the same variables used to construct `object`.

The output, with class `margins`, is a list that contains only the vector margins. Next, we show the code to calculate the margins of the previously built bagging and boosting classifiers for the training and test sets. Examples with negative margins are those that have been incorrectly classified. Both classifiers commit 5 and 1 errors for the training and test sets, respectively.

```
> # Training set
> margins(iris.bagging, iris[train,])->iris.bagging.margins # bagging
> sort(margins(iris.bagging,iris[train,])$marg)

 [1] -0.4 -0.4 -0.2 -0.2 -0.2  0.2  0.2  0.2  0.2  0.2  0.2  0.2  0.2  0.2  0.2
[16]  0.2  0.2  0.2  0.2  0.2  0.2  0.2  0.2  0.2  0.2  0.2  0.2  0.2  0.2  0.2
[31]  0.2  0.2  0.2  0.2  0.2  0.2  0.2  0.2  0.2  0.2  0.2  0.4  0.4  0.4  0.4
[46]  0.4  0.4  0.4  0.4  0.4  0.4  0.4  0.4  0.4  0.4  0.4  0.4  0.4  0.4  0.4
[61]  0.4  0.4  0.4  0.4  0.4  0.4  0.4  0.4  0.4  0.4  0.4  0.4  0.4  0.4  0.4
[76]  0.4  0.4  0.4  0.4  0.4  0.4  0.4  0.4  0.4  0.4  0.4  0.4  0.4  0.4  0.4
[91]  0.4  0.4  0.4  0.4  0.4  0.4  0.4  0.4  0.4

> margins(iris.adaboost,iris[train,])->iris.adaboost.margins # boosting
> round( sort(margins(iris.adaboost,iris[train,])$marg), digits=3)

 [1] -0.285 -0.285 -0.285 -0.191 -0.076  0.025  0.076  0.076  0.076  0.242  0.242
[12]  0.242  0.242  0.242  0.242  0.242  0.242  0.242  0.242  0.242  0.242  0.242
[23]  0.242  0.242  0.242  0.242  0.242  0.242  0.242  0.242  0.242  0.242  0.242
[34]  0.242  0.242  0.285  0.306  0.306  0.306  0.306  0.306  0.306  0.306  0.306
[45]  0.306  0.306  0.306  0.306  0.306  0.306  0.306  0.306  0.306  0.306  0.306
[56]  0.306  0.306  0.306  0.306  0.306  0.306  0.306  0.306  0.306  0.306  0.306
[67]  0.306  0.306  0.306  0.452  0.452  0.452  0.452  0.452  0.452  0.452  0.452
[78]  0.452  0.452  0.452  0.452  0.452  0.452  0.452  0.452  0.452  0.452  0.452
[89]  0.452  0.452  0.452  0.452  0.452  0.452  0.452  0.452  0.452  0.452  0.452

> # Test set
> margins(iris.predbagging, iris[-train,])->iris.predbag.margins # bagging
> sort(margins(iris.predbagging,iris[-train,])$marg)

 [1] -0.2  0.2  0.2  0.2  0.2  0.2  0.2  0.2  0.2  0.2  0.2  0.2  0.2  0.2  0.2  0.2
[17]  0.2  0.2  0.2  0.2  0.4  0.4  0.4  0.4  0.4  0.4  0.4  0.4  0.4  0.4  0.4  0.4
[33]  0.4  0.4  0.4  0.4  0.4  0.4  0.4  0.4  0.4  0.4  0.4  0.4  0.4  0.4  0.4  0.4
[49]  0.4  0.4  0.4

> margins(iris.predboosting,iris[-train,])->iris.predboost.margins # boosting
> sort(margins(iris.predboosting, iris[-train,])$marg)

 [1] -0.28544296  0.07593561  0.07593561  0.24209719  0.24209719  0.24209719
 [7]  0.24209719  0.24209719  0.24209719  0.24209719  0.24209719  0.24209719
[13]  0.24209719  0.24209719  0.24209719  0.24209719  0.24209719  0.30629828
[19]  0.30629828  0.30629828  0.30629828  0.30629828  0.30629828  0.30629828
[25]  0.30629828  0.30629828  0.30629828  0.30629828  0.30629828  0.30629828
[31]  0.30629828  0.30629828  0.30629828  0.30629828  0.45160454  0.45160454
[37]  0.45160454  0.45160454  0.45160454  0.45160454  0.45160454  0.45160454
[43]  0.45160454  0.45160454  0.45160454  0.45160454  0.45160454  0.45160454
[49]  0.45160454  0.45160454  0.45160454
```

Another plot available since version 4.0 of **adabag** is `plot.margins()`. This function represents the previously calculated margins of an AdaBoost.M1, AdaBoost-SAMME, or bagging classifier for a data frame. It has the same arguments as the previous plot, that is, an object x of class `margins`, with a component named

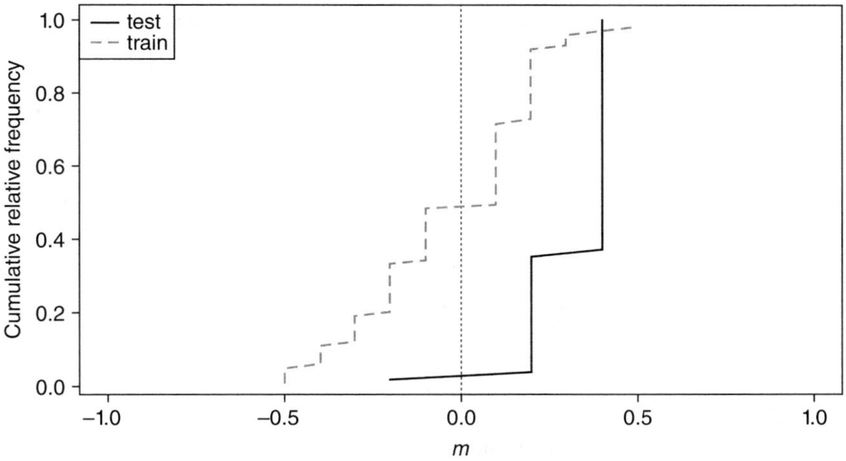

Figure 4.5 Margins for bagging in the iris example.

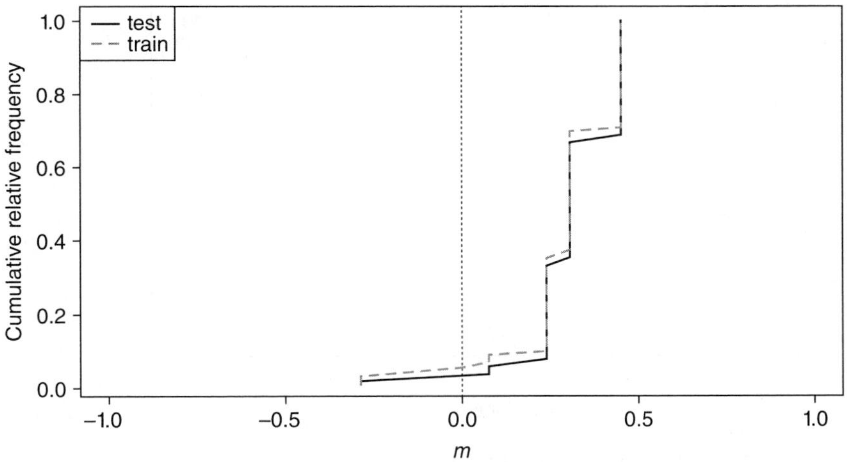

Figure 4.6 Margins for boosting in the iris example.

margins, as that returned by the `margins` function. The second argument *y* can be used to represent in the same plot the margins in the test and training sets, *x* and *y*, respectively. It should be NULL (by default) or an object of class margins. The three dots have the same meaning. Intuitively, the margin for an observation is related to the certainty of its classification. It is calculated as the difference between the support of the correct class and the maximum support of an incorrect class. The following code returns Figures 4.5 and 4.6.

```
> plot.margins(iris.predbag.margins, iris.bagging.margins)
> plot.margins(iris.predboost.margins, iris.adaboost.margins)
```

With regard to the `errorevol` function of the **adabag** library, this calculates the evolution of the classifier error AdaBoost.M1, AdaBoost-SAMME or bagging for a data set as the size of the combination grows. This function has the same two arguments as

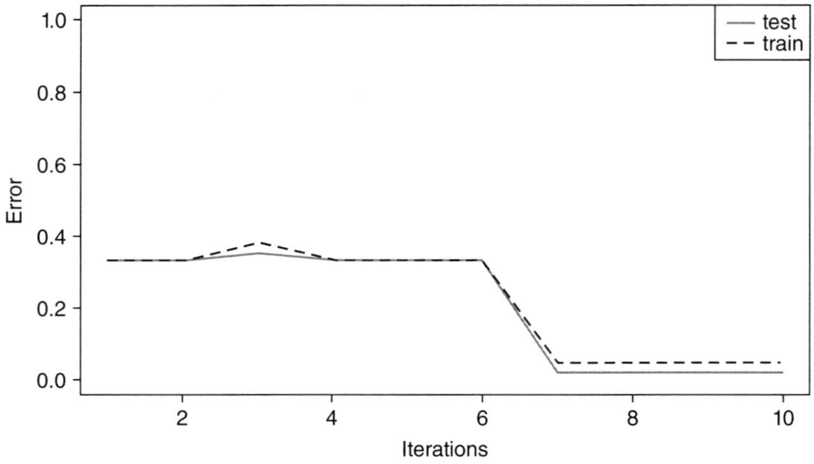

Figure 4.7 Error evolution in bagging for the iris example.

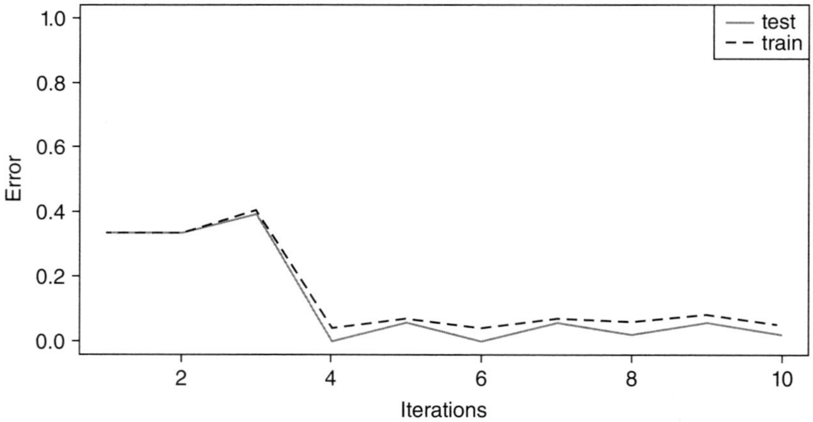

Figure 4.8 Error evolution in boosting for the iris example.

`margins` and returns a list with the vector with the evolution of the error. This can be used to see the speed with which bagging or boosting reduces the error of the combination. In addition, it would help us to detect the presence of overfitting and, in that case, the convenience of pruning the combination using the corresponding `predict` function. Although this example is not the most appropriate case to see the usefulness of the error evolution, due to the small size of the combination, we use it only for didactic purposes. In the case of bagging the number of trees could be reduced to six and for boosting four trees achieve a better result than the whole ensemble. Figures 4.7 and 4.8 show these evolutions graphically.

```
> # Error evolution for bagging in training and test set
> errorevol(iris.bagging,iris[train,])-> bagg.evol.train
> bagg.evol.train

$error
[1] 0.33333333 0.33333333 0.33333333 0.33333333 0.33333333 0.05050505
```

```
[7] 0.05050505 0.05050505 0.05050505 0.05050505
attr(,"class")
[1] "errorevol"

> errorevol(iris.bagging,iris[-train,])-> bagg.evol.test
> bagg.evol.test$err

[1] 0.33333333 0.33333333 0.33333333 0.33333333 0.33333333 0.01960784
[7] 0.01960784 0.01960784 0.01960784 0.01960784

> # Error evolution for boosting in training and test set
> errorevol(iris.adaboost,iris[train,])-> boost.evol.train
> boost.evol.train$err

 $error
 [1] 0.33333333 0.33333333 0.40404040 0.04040404 0.07070707 0.04040404
 [7] 0.07070707 0.06060606 0.08080808 0.05050505

> errorevol(iris.adaboost,iris[-train,])-> boost.evol.test
> boost.evol.test$err

[1] 0.33333333 0.33333333 0.39215686 0.00000000 0.05882353 0.00000000
[7] 0.05882353 0.01960784 0.05882353 0.01960784
```

The third plot function is `plot.errorevol`, which draws the previously calculated error evolution of an AdaBoost.M1, AdaBoost-SAMME or bagging classifier for a data frame as the ensemble size grows. It has three initial arguments. First, *x*, an object of class `errorevol`. This is assumed to be the result of the same function that produces an object with a component named `error` as that returned by the `errorevol` function. The second argument *y* can be used to represent in the same plot the evolution of the test and training errors, *x* and *y*, respectively. This should be NULL (by default) or an object of class `errorevol`. Finally, the three dots enable the use of further arguments passed to or from other methods.

This kind of plot can be useful to see how fast bagging or boosting reduces the error of the ensemble. In addition, it can detect the presence of overfitting and, therefore, the convenience of pruning the ensemble using `predict.bagging` or `predict.boosting`. A labeled plot is produced on the current graphics device (one being opened if needed).

Figures 4.7 and 4.8 show error evolution for bagging and boosting using the plot.errorevol() function.

```
> plot.errorevol(bagg.evol.test, bagg.evol.train, main="Bagging error vs
number of trees")
> plot.errorevol(boost.evol.test, boost.evol.train, main="Boosting error vs
number of trees")
```

4.2.4 The `MarginOrderedPruning.Bagging` Function

The `MarginOrderedPruning.Bagging` function allows us to use margin-based ordered aggregation for bagging pruning (Guo and Boukir, 2013). To apply this function, we have the following five arguments:

- `baggingObject`: a fitted model object of class bagging.
- `trainingset`: the training set of the bagging object.
- `pruningset`: a set-aside data set for bagging pruning.

- marginType: if "unsupervised" (by default) the margin is the difference between the proportions of votes of the first and second most popular classes. Otherwise the margin is calculated as the difference between the proportion of votes of the correct class and the most popular among the other classes.
- doTrace: if set to TRUE, give a more verbose output as MarginOrderedPruning.Bagging is running.

The output returns a list with the following components:

- prunedBagging: a pruned bagging object that can be used for predicting as it has class bagging.
- AccuracyOrderedEnsemblePruningSet: the accuracy of each ordered ensemble on the pruning set.

Taking again the example of iris, in this case we separate the observations into three sets: training, pruning, and test. The random sample results have 45, 31 and 74 observations, respectively, for each of these sets. We maintain the size of the trees as decision stumps, but increase their number to 50. The test error that achieves bagging with 50 trees is 4.05%, while the combination pruned by the margin-based ordered method achieves the same test error using only 15 trees, as shown below.

```
## Separate data into 3 parts: training set, pruning set and test set
>   ind <- sample(3, nrow(iris), replace = TRUE, prob=c(0.3, 0.2,0.5))
> length(which(ind==1))
[1] 45
> length(which(ind==2))
[1] 31
> length(which(ind==3))
[1] 74

## create bagging with training set and apply it in the test set
> iris.bagging1<-bagging(Species~.,data=iris[ind==1,],mfinal=50,
 control=rpart.control(maxdepth=1))
>   iris.bagging.pred<-predict(iris.bagging1,iris[ind==3,])
>   iris.bagging.pred$conf

               Observed Class
Predicted Class setosa versicolor virginica
     setosa         24          0         0
     versicolor      0         27         2
     virginica       0          1        20
>   iris.bagging.pred$error
[1] 0.04054054

>   ##pruning bagging
>   iris.bagging.pruning<-MarginOrderedPruning.Bagging(iris.bagging1,
+      iris[ind==1,],iris[ind==2,])

> iris.bagging.pruning.pred<-predict(iris.bagging.pruning$prunedBagging,
iris[ind==3,])
>   iris.bagging.pruning.pred$confusion

               Observed Class
Predicted Class setosa versicolor virginica
     setosa         24          0         0
     versicolor      0         28         3
     virginica       0          0        19
>   iris.bagging.pruning.pred$error
[1] 0.04054054
```

```
> iris.bagging.pruning

$prunedBagging
$formula
Species ~.

$trees
$trees[[1]]
n= 45

node), split, n, loss, yval, (yprob)
      * denotes terminal node

1) root 45 28 versicolor (0.2666667 0.3777778 0.3555556)
  2) Petal.Length< 2.7 12  0 setosa (1.0000000 0.0000000 0.0000000) *
  3) Petal.Length>=2.7 33 16 versicolor (0.0000000 0.5151515 0.4848485) *

$trees[[2]]
n= 45

node), split, n, loss, yval, (yprob)
      * denotes terminal node

1) root 45 28 virginica (0.2666667 0.3555556 0.3777778)
  2) Petal.Length< 2.6 12  0 setosa (1.0000000 0.0000000 0.0000000) *
  3) Petal.Length>=2.6 33 16 virginica (0.0000000 0.4848485 0.5151515) *

$trees[[3]]
n= 45

node), split, n, loss, yval, (yprob)
      * denotes terminal node

1) root 45 25 versicolor (0.2444444 0.4444444 0.3111111)
  2) Petal.Length< 2.6 11  0 setosa (1.0000000 0.0000000 0.0000000) *
  3) Petal.Length>=2.6 34 14 versicolor (0.0000000 0.5882353 0.4117647) *

$trees[[4]]
n= 45

node), split, n, loss, yval, (yprob)
      * denotes terminal node

1) root 45 23 setosa (0.4888889 0.2222222 0.2888889)
  2) Petal.Length< 2.5 22  0 setosa (1.0000000 0.0000000 0.0000000) *
  3) Petal.Length>=2.5 23 10 virginica (0.0000000 0.4347826 0.5652174) *

$trees[[5]]
n= 45

node), split, n, loss, yval, (yprob)
      * denotes terminal node

1) root 45 24 virginica (0.24444444 0.28888889 0.46666667)
  2) Petal.Length< 4.65 22 11 setosa (0.50000000 0.50000000 0.00000000) *
  3) Petal.Length>=4.65 23  2 virginica (0.00000000 0.08695652 0.91304348) *

$trees[[6]]
n= 45

node), split, n, loss, yval, (yprob)
      * denotes terminal node
```

```
1) root 45 28 virginica (0.2666667 0.3555556 0.3777778)
  2) Petal.Length< 2.6 12  0 setosa (1.0000000 0.0000000 0.0000000) *
  3) Petal.Length>=2.6 33 16 virginica (0.0000000 0.4848485 0.5151515) *

$trees[[7]]
n= 45

node), split, n, loss, yval, (yprob)
      * denotes terminal node

1) root 45 27 versicolor (0.2666667 0.4000000 0.3333333)
  2) Petal.Length< 2.7 12  0 setosa (1.0000000 0.0000000 0.0000000) *
  3) Petal.Length>=2.7 33 15 versicolor (0.0000000 0.5454545 0.4545455) *

$trees[[8]]
n= 45

node), split, n, loss, yval, (yprob)
      * denotes terminal node

1) root 45 29 versicolor (0.3111111 0.3555556 0.3333333)
  2) Petal.Length< 2.6 14  0 setosa (1.0000000 0.0000000 0.0000000) *
  3) Petal.Length>=2.6 31 15 versicolor (0.0000000 0.5161290 0.4838710) *

$trees[[9]]
n= 45

node), split, n, loss, yval, (yprob)
      * denotes terminal node

1) root 45 28 versicolor (0.2888889 0.3777778 0.3333333)
  2) Petal.Length< 2.55 13  0 setosa (1.0000000 0.0000000 0.0000000) *
  3) Petal.Length>=2.55 32 15 versicolor (0.0000000 0.5312500 0.4687500) *

$trees[[10]]
n= 45

node), split, n, loss, yval, (yprob)
      * denotes terminal node

1) root 45 28 versicolor (0.2666667 0.3777778 0.3555556)
  2) Petal.Length< 2.5 12  0 setosa (1.0000000 0.0000000 0.0000000) *
  3) Petal.Length>=2.5 33 16 versicolor (0.0000000 0.5151515 0.4848485) *

$trees[[11]]
n= 45

node), split, n, loss, yval, (yprob)
      * denotes terminal node

1) root 45 25 versicolor (0.2222222 0.4444444 0.3333333)
  2) Petal.Length< 2.7 10  0 setosa (1.0000000 0.0000000 0.0000000) *
  3) Petal.Length>=2.7 35 15 versicolor (0.0000000 0.5714286 0.4285714) *

$trees[[12]]
n= 45

node), split, n, loss, yval, (yprob)
      * denotes terminal node

1) root 45 25 virginica (0.2000000 0.3555556 0.4444444)
  2) Petal.Width< 1.6 28 12 versicolor (0.3214286 0.5714286 0.1071429) *
```

```
  3) Petal.Width>=1.6 17  0 virginica (0.0000000 0.0000000 1.0000000) *

$trees[[13]]
n= 45

node), split, n, loss, yval, (yprob)
      * denotes terminal node

1) root 45 20 virginica (0.2444444 0.2000000 0.5555556)
  2) Petal.Width< 1.6 20  9 setosa (0.5500000 0.4500000 0.0000000) *
  3) Petal.Width>=1.6 25  0 virginica (0.0000000 0.0000000 1.0000000) *

$trees[[14]]
n= 45

node), split, n, loss, yval, (yprob)
      * denotes terminal node

1) root 45 27 virginica (0.26666667 0.33333333 0.40000000)
  2) Petal.Width< 1.75 28 13 versicolor (0.42857143 0.53571429 0.03571429) *
  3) Petal.Width>=1.75 17  0 virginica (0.00000000 0.00000000 1.00000000) *

$trees[[15]]
n= 45

node), split, n, loss, yval, (yprob)
      * denotes terminal node

1) root 45 24 virginica (0.2444444 0.2888889 0.4666667)
  2) Petal.Length< 4.75 25 12 versicolor (0.4400000 0.5200000 0.0400000) *
  3) Petal.Length>=4.75 20  0 virginica (0.0000000 0.0000000 1.0000000) *

attr(,"vardep.summary")
    setosa versicolor  virginica
        13         15         17
attr(,"class")
[1] "bagging"

$AccuracyOrderedEnsemblePruningSet
 [1] 0.6451613 0.6451613 0.6451613 0.6451613 0.9354839 0.7741935 0.9354839 0.6451613
 [9] 0.6451613 0.6451613 0.6451613 0.6451613 0.6451613 0.6451613 1.0000000 1.0000000
[17] 1.0000000 1.0000000 1.0000000 1.0000000 0.9677419 1.0000000 0.9677419 0.9677419
[25] 0.9677419 0.9677419 0.9677419 0.9677419 0.9677419 1.0000000 0.9677419 1.0000000
[33] 1.0000000 1.0000000 0.9677419 1.0000000 0.9677419 1.0000000 1.0000000 1.0000000
[41] 1.0000000 1.0000000 1.0000000 1.0000000 1.0000000 1.0000000 1.0000000 1.0000000
[49] 1.0000000 1.0000000
```

4.3 The "German Credit" Example

In this section, we will further illustrate the classification by trees and the combination procedures of trees, using a more complex example, "German Credit", which also proceeds from the UCI Machine Learning Repository (Dua and Karra Taniskidou, 2017).

The task is to classify clients as "good" or "bad" when granting a loan based on the risk they have according to certain variables. The file contains 1000 observations and 21 variables, 7 of which are quantitative and 14 categorical, including the class. The first thing we do is read the text file "german.data" downloaded from the UCI database http://archive.ics.uci.edu/ml/index.php and define the attributes as factors. We then assign them the categories that appear in the description file attached in UCI. We can observe that the last factor is the class with two categories "Good" and "Bad".

```
> german = read.table("german.data", header=F)
> dim(german)
> names(german) <- c("checking.account", "duration", "credit.history", "purpose",
+ "credit.amount","savings.account","employment","installment.rate","personal.status",
+ "debtors","residence","property","age","other.installment.plans","housing",
+ "existing.credits","job","maintenance.for","telephone","foreign.worker","class")
> german$class <- as.factor(german$class)
>
> levels(german$checking.account) <- c("...< 0 DM", "0 <=...<
+ 200 DM","...>= 200 DM",
+ "no checking account")
> levels(german$credit.history) <- c("no credits taken/all credits paid back duly",
+ "all credits at this bank paid back duly", "existing credits paid back duly till
now",
+ "delay in paying off in the past", "critical account/other credits existing")
> levels(german$purpose) <- c("car(new)", "car(used)", "furniture/equipment",
+ "radio/television", "domestic appliances",
+ "repairs", "education", "vacation - does not exist?", "retraining", "business",
+ "others")
> levels(german$savings.account) <- c("... <  100 DM", "100 <= ... <  500 DM",
+ "500 <= ... < 1000 DM", "... >= 1000 DM", "unknown/ no savings account")
> levels(german$employment) <- c("unemployed", "... < 1 year", "1 <= ... < 4 years",
+ "4  <= ... < 7 years", "... >= 7 years")
> levels(german$personal.status) <- c("male: divorced/separated",
+ "female: divorced/separated/married", "male: single", "male: married/widowed",
+ "female: single")
> levels(german$debtors) <- c("none", "co-applicant", "guarantor")
> levels(german$property) <- c("real estate",
+ "if not: building society savings agreement/life insurance", "if not: car or
other",
+ "unknown / no property")
> levels(german$other.installment.plans) <- c("bank", "stores", "none")
> levels(german$housing) <- c("rent", "own", "for free")
> levels(german$job) <- c("unemployed/ unskilled  - non-resident",
"unskilled-resident",
+ "skilled employee / official",
+ "management/ self-employed/highly qualified employee/ officer")
> levels(german$telephone) <- c("none", "yes")
> levels(german$foreign.worker) <- c("yes", "no")
> levels(german$class) <- c("Good","Bad")
```

We perform a basic statistic of all the variables and we see that there are 700 individuals classified as "Good" and 300 as "Bad". We then randomly select 75% of the individuals as a training set, whose index vector is called train, and the rest are reserved as the test set.

```
> summary(german)

            checking.account     duration
 ...< 0 DM            :274     Min.   : 4.0
 0 <=...< 200 DM      :269     1st Qu.:12.0
 ...>= 200 DM         : 63     Median :18.0
 no checking account:394       Mean   :20.9
                               3rd Qu.:24.0
                               Max.   :72.0

                                      credit.history                     purpose
 no credits taken/all credits paid back duly: 40    domestic appliances        :280
```

```
all credits at this bank paid back duly       : 49    car(new)                      :234
existing credits paid back duly till now      :530    radio/television              :181
delay in paying off in the past               : 88    car(used)                     :103
critical account/other credits existing       :293    business                      : 97
                                                      vacation - does not exist?: 50
                                                      (Other)                       : 55
credit.amount                          savings.account                employment
Min.    :  250    ... <  100 DM                :603    unemployed             : 62
1st Qu.: 1366    100 <= ... <  500 DM          :103    ... < 1 year           :172
Median : 2320    500 <= ... < 1000 DM          : 63    1  <= ... < 4 years:339
Mean   : 3271    ... >= 1000 DM                : 48    4  <= ... < 7 years:174
3rd Qu.: 3972    unknown/ no savings account:183       ... >= 7 years         :253
Max.   :18424

installment.rate                             personal.status          debtors
Min.   :1.000    male  : divorced/separated          : 50    none          :907
1st Qu.:2.000    female: divorced/separated/married :310    co-applicant: 41
Median :3.000    male  : single                      :548    guarantor    : 52
Mean   :2.973    male  : married/widowed             : 92
3rd Qu.:4.000    female: single                      :  0
Max.   :4.000
   residence                                                          property
Min.   :1.000    real estate                                              :282
1st Qu.:2.000    if not: building society savings agreement/life insurance:232
Median :3.000    if not: car or other                                     :332
Mean   :2.845    unknown / no property                                    :154
3rd Qu.:4.000
Max.   :4.000

     age          other.installment.plans      housing       existing.credits
Min.   :19.00    bank:139                   rent     :179    Min.   :1.000
1st Qu.:27.00    stores: 47                 own      :713    1st Qu.:1.000
Median :33.00    none:814                   for free:108    Median :1.000
Mean   :35.55                                                Mean   :1.407
3rd Qu.:42.00                                                3rd Qu.:2.000
Max.   :75.00                                                Max.   :4.000

                                                                  job        maintenance.for
unemployed/ unskilled  - non-resident                              : 22      Min.   :1.000
unskilled - resident                                               :200      1st Qu.:1.000
skilled employee / official                                        :630      Median :1.000
management/ self-employed/highly qualified employee/ officer:148      Mean   :1.155
                                                                             3rd Qu.:1.000
                                                                             Max.   :2.000

telephone   foreign.worker   class
none:596    yes:963          Good:700
yes:404     no: 37           Bad:300

> train<- sample(1:1000, 750)
```

4.3.1 Classification Tree

We load the recursive partitioning library **rpart** that allows us to develop classification trees. The pruning of the tree can be done manually, from the table of cost complexity, or we can use the previously mentioned `autoprune` function in the **adabag** package, which is based on the rule of pruning the tree at the point at which the minimum error of cross-validation plus a standard deviation is reached.

We then estimate the tree in the training set to explain the class as a function of the rest of the variables. We must remember that it is not necessary to select variables, the trees will only use those that allow a better discrimination between the categories of the class.

We start developing the tree completely, i.e., achieving a 0% error in the training set, which triggers the error in the test set to 30.4%. This is seen in the corresponding matrices of confusion and in the errors calculated from them.

```
> formula <- class~.
> german.rpart <- rpart(formula, data=german, subset=train,method="class",
+ parms=list(split="information"),
+ control=rpart.control(cp=0,xval=10,maxdepth=30,minsplit=2,minbucket=
+ 1))
> #training set
> german.predrpart<-predict(german.rpart, newdata = german[train,], type="class")
> #Confusion matrix
> table(german.predrpart, vardep[train], dnn=c("Predicted class", "Observed class"))

               Observed class
Predicted class Good Bad
           Good  536   0
           Bad     0 214

> #Error
> 1-sum(german.predrpart==vardep[train])/length(vardep[train])

[1] 0

> # Test set
> german.predrpart<-predict(german.rpart, newdata = german[-train,], type="class")
> #Confusion matrix
> table(german.predrpart, vardep[-train], dnn=c("Predicted class",
"Observed class"))

               Observed class
Predicted class Good Bad
           Good  137  49
           Bad    27  37

> #Error
> 1-sum(german.predrpart==vardep[-train])/length(vardep[-train])

[1] 0.304
```

Figure 4.9 represents the fully developed tree (overfitted) and we can observe the enormous complexity needed to correctly classify all the individuals in the training set. To obtain a tree with the errors in both sets more balanced and, therefore, with higher generalization capacity, we prune it with the function `autoprune`.

```
> german.prune<-autoprune(formula, german, train)
> german.prune

[[1]]
n= 750

node), split, n, loss, yval, (yprob)
      * denotes terminal node

 1) root 750 214 Good (0.7146667 0.2853333)
   2) checking.account=...>= 200 DM,no checking account 353  45 Good (0.8725212
0.1274788) *
   3) checking.account=...< 0 DM,0 <=...< 200 DM 397 169 Good (0.5743073 0.4256927)
```

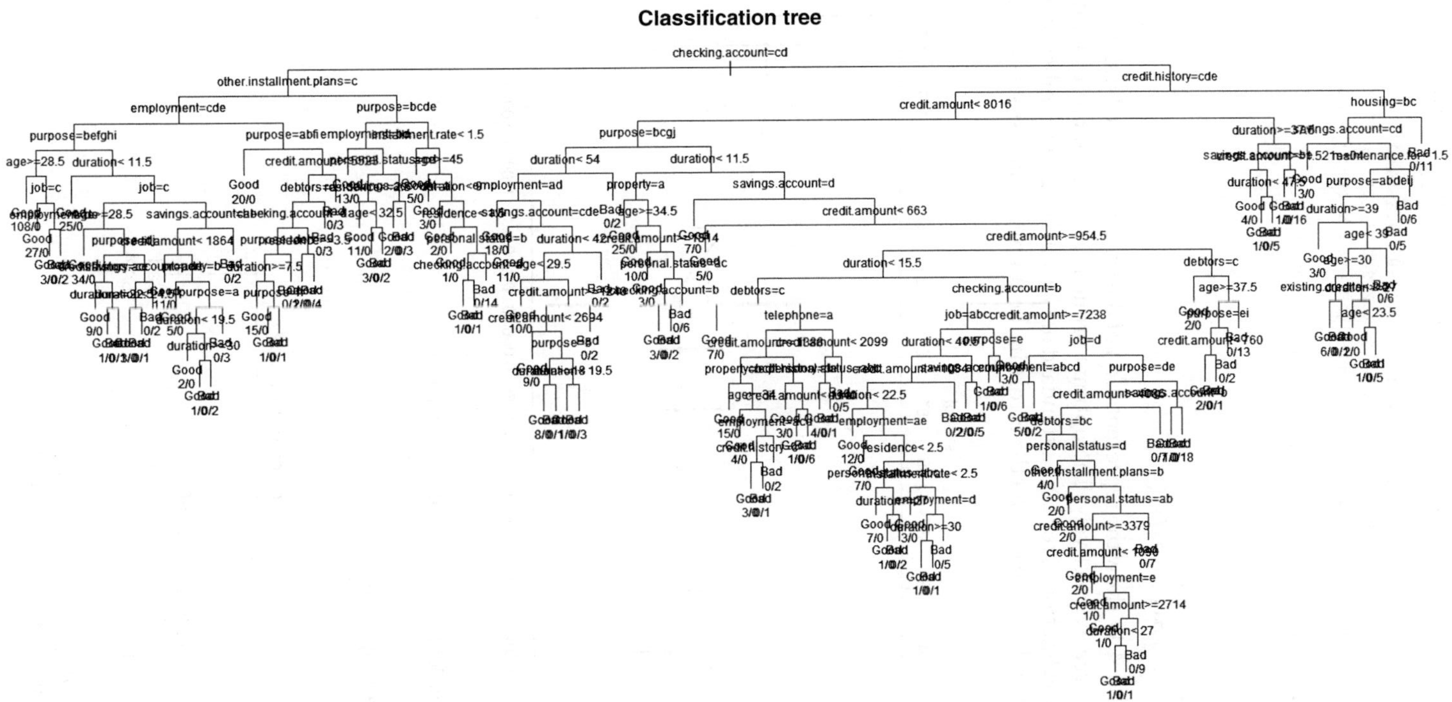

Figure 4.9 Overfitted classification tree.

```
      6) credit.history=existing credits paid back duly till now,delay in paying off
      in the past,critical account/other credits existing 348 135 Good (0.6120690
0.3879310)
        12) credit.amount< 8015.5 321 114 Good (0.6448598 0.3551402) *
        13) credit.amount>=8015.5 27   6 Bad (0.2222222 0.7777778) *
      7) credit.history=no credits taken/all credits paid back duly,all cred-
its at this
      bank paid back duly 49  15 Bad (0.3061224 0.6938776) *

[[2]]
[1] 0.01518692

> cp.prune <- german.prune[[2]]
> german.prune <- german.prune[[1]]
```

Figures 4.10 and 4.11 show the representation of the cp table versus the size of the tree and the pruned tree, respectively. The optimal pruned tree, which corresponds to the minimum cross-validation error plus a standard deviation, is size 4 (number of terminal nodes or leaves) and it has balanced the errors in the training and test sets at 24.0% and

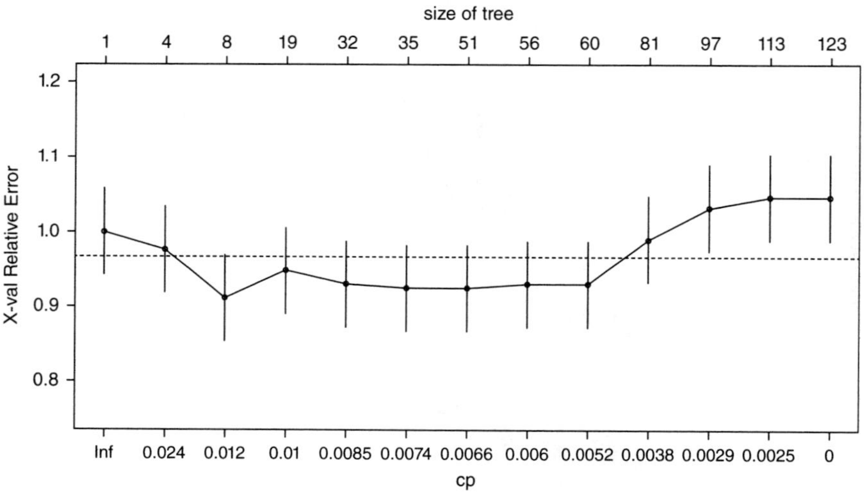

Figure 4.10 Cross-validation error versus tree complexity.

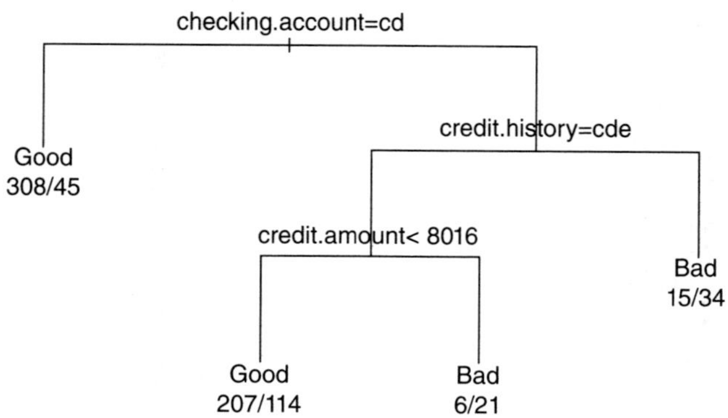

Figure 4.11 Pruned tree.

30.8%, respectively. We see that the errors in the test set are similar, but the latter tree is preferable to the first one, in addition to the reason given above, because following the "parsimony principle" or Ockham's razor, a simpler model is preferable.

```
> # Training set
> german.predprune<-predict(german.prune, newdata = german[train,], type="class")
> table(german.predprune, vardep[train], dnn=c("Predicted class", "Observed class"))

               Observed class
Predicted class Good Bad
           Good  515 159
           Bad    21  55

> 1-sum(german.predprune==vardep[train])/length(vardep[train])

[1] 0.24

> # Test set
>  german.predprune<-predict(german.prune, newdata = german[-train,], type="class")
>  table(german.predprune, vardep[-train], dnn=c("Predicted class",
"Observed class"))

               Observed class
Predicted class Good Bad
           Good  155  68
           Bad     9  18

>  1-sum(german.predprune==vardep[-train])/length(vardep[-train])

[1] 0.308

> plot(german.prune, uniform=T, margin=0.1)
> title( main="Pruned tree", cex.main=3)
> text(german.prune, use.n=TRUE, cex=2)
```

4.3.2 Combination using Bagging

Once the individual classification tree has been constructed we will try to improve its results by combining trees of the same size or even smaller, applying the algorithms bagging and AdaBoost.M1 described in the previous section. We do not apply SAMME since it is equivalent in the dichotomous case, which is the one that occupies us. We start by bagging using the same training and test sets as in the previous tree, with a maximum depth of 3 for the individual trees that make up the combination and a maximum of 200 trees in it that we can later prune.

```
>  control <- rpart.control(maxdepth = 3, minsplit = 2, cp = -1, xval=0)
> mfinal <- 200
> german.bagging <- bagging(formula=formula, data=german[train,], mfinal=mfinal,
+ control=control)
>
> #Training set
> german.predbag <- predict.bagging(german.bagging, newdata=german[train,])
> german.predbag$confusion

               Observed Class
Predicted Class Good Bad
            Bad    7  56
            Good 529 158

> german.predbag$error

[1] 0.22
```

```
> #Test set
> german.predbag <- predict.bagging(german.bagging, newdata=german[-train,])
> german.predbag$confusion

               Observed Class
Predicted Class Good Bad
           Bad     3  18
           Good  161  68

> german.predbag$error

[1] 0.284

> sort(german.bagging$importance,dec=T)

       checking.account                  purpose                 duration
            42.12377430              11.31422108               8.86604391
          credit.history           credit.amount other.installment.plans
             8.01114993               7.54283410               7.04972877
        savings.account               employment                      age
             3.64336605               3.58020228               2.69401470
               property                  debtors          personal.status
             1.61318710               0.94658312               0.80922474
                    job         installment.rate                residence
             0.68498935               0.41910285               0.38487963
                housing         existing.credits           foreign.worker
             0.25226528               0.06443282               0.00000000
        maintenance.for                telephone
             0.0                            0.00000000

> importanceplot(german.bagging, horiz=F, cex.names=.9)
```

As we can see, the errors have decreased 2 and 2.4 points in the training and test sets with respect to the individual tree, which means an improvement of approximately 8%. The most discriminative variables, ordered according to their importance in Figure 4.12, are "checking account" (42.12%) and "purpose" (11.3%).[2] The estimate can still be improved by pruning the combination. Observing the evolution of the error (Figure 4.13), we stop at that point in which the error in the test set is minimal, since from that point overfitting occurs and, therefore, the error increases.

```
> # Error evolution
> errorevol(german.bagging, german[train,]) -> errorevol.train
> errorevol(german.bagging, german[-train,]) -> errorevol.test
>
> plot.errorevol(errorevol.test, errorevol.train, main="Bagging error vs
number of trees")
> abline(h=min(errorevol.test[[1]]), col="red",lty=3)
> abline(h=min(errorevol.train[[1]]), col="blue",lty=3)

> newmfinal = which(errorevol.test[[1]]==min(errorevol.test[[1]]))
> newmfinal[1]

[1] 51

> errorevol.train$error[51]

[1] 0.2093333

> errorevol.test$error[51]

[1] 0.26
```

2 The following codification is used in some graphics for better visibility: V1, age; V2, checking.account; V3, credit.amount; V4, credit.history; V5, debtors; V6, duration; V7, employment; V8, existing.credits; V9, foreign.worker; V10, housing; V11, installment.rate; V12, job; V13, maintenance.for; V14, other.installment.plans; V15, personal.status; V16, property; V17, purpose; V18, residence; V19, savings.account; V20, telephone.

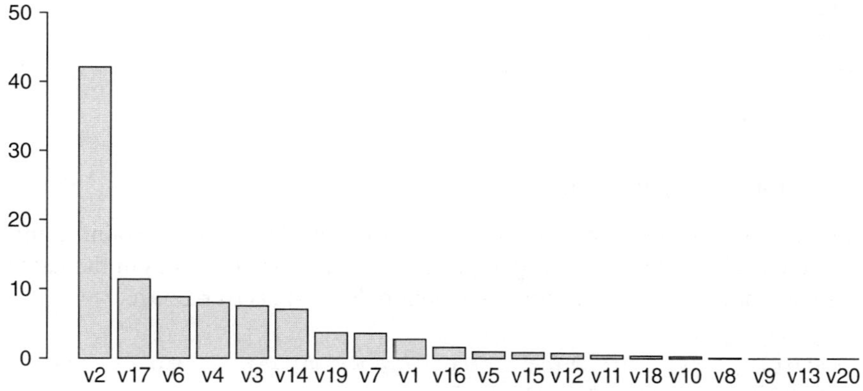

Figure 4.12 Variable relative importance in bagging.

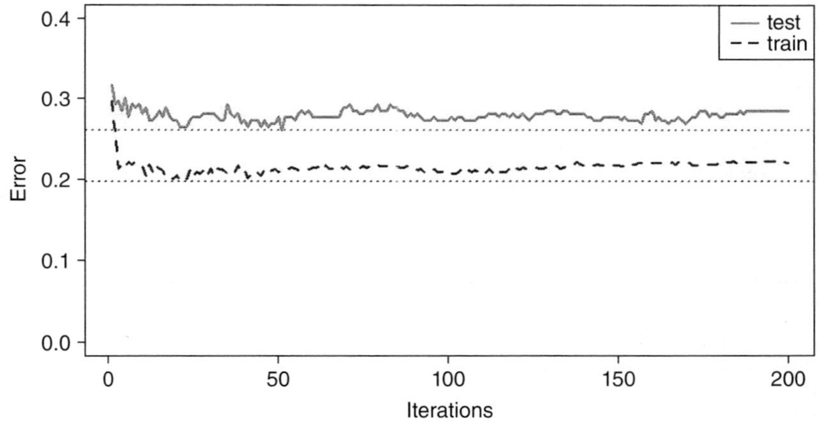

Figure 4.13 Error evolution in bagging.

The minimum error in the test set is obtained for a combination of 51 trees and the errors obtained in training and testing for bagging pruned to 51 trees are 20.9% and 26%, which has substantially improved the results of both the individual tree and the bagging combination without pruning. The confusion matrices for both sets are shown below.

```
> #Training set
> german.predbag <- predict.bagging(german.bagging, newdata=german[train,],
+ newmfinal=newmfinal[1])
> german.predbag$confusion

               Observed Class
Predicted Class Good Bad
          Bad     12  69
          Good   524 145

> german.predbag$error

[1] 0.2093333

> #Test set
> german.predbag <- predict.bagging(german.bagging, newdata=german[-train,],
+ newmfinal=newmfinal[1])
> german.predbag$confusion

               Observed Class
Predicted Class Good Bad
```

```
           Bad      4  25
           Good   160  61

> german.predbag$error

[1] 0.26
```

4.3.3 Combination using Boosting

To combine individual classifiers by the boosting method, we use the same training and test sets as in the individual tree and bagging, and the same parameters as in the latter (200 trees of maximum depth 3) and the coefficient of learning is `Breiman`.

```
> coeflearn <- "Breiman"
> german.boosting <- boosting(formula=formula, data=german[train,], mfinal=mfinal,
+ boos=F,control=control, coeflearn=coeflearn)
>
> #Training set
> german.predboost<- predict.boosting(german.boosting, newdata=german[train,])
> german.predboost$confusion

               Observed Class
Predicted Class Good Bad
           Bad    18 177
           Good  518  37

> german.predboost$error

[1] 0.07333333

> #Test set
> german.predboost<- predict.boosting(german.boosting, newdata=german[-train,])
> german.predboost$confusion

               Observed Class
Predicted Class Good Bad
           Bad    16  37
           Good  148  49

> german.predboost$error

[1] 0.26

> sort(german.boosting$importance,dec=T)

       checking.account              credit.amount                    purpose
            22.56077221                18.37466644                11.78889167
               duration other.installment.plans                       age
            11.03844408                 5.53445585                 4.89833937
        savings.account             credit.history                 employment
             4.85173233                 4.51882958                 3.63439071
               property                    debtors                  residence
             2.31837963                 2.12855608                 1.83401272
        personal.status                        job             foreign.worker
             1.78156649                 1.56803904                 1.19527395
       installment.rate                    housing                  telephone
             1.11437921                 0.34924967                 0.23268643
       existing.credits            maintenance.for
             0.19036371                 0.08697084
```

As we can see, errors have dropped considerably, especially in the training set. The reduction is 17 and 4.2 points in the training and test sets with respect to the individual tree, and 15 and 2.4 points with respect to bagging. The importance of the most discriminating variables is more distributed than in bagging and, according to their importance

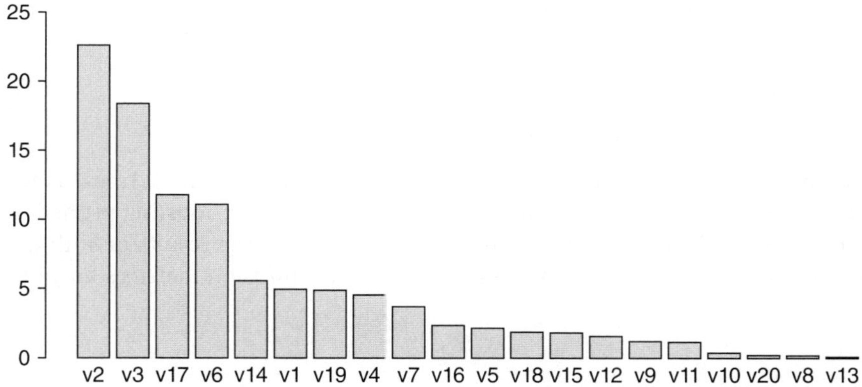

Figure 4.14 Variable relative importance in boosting.

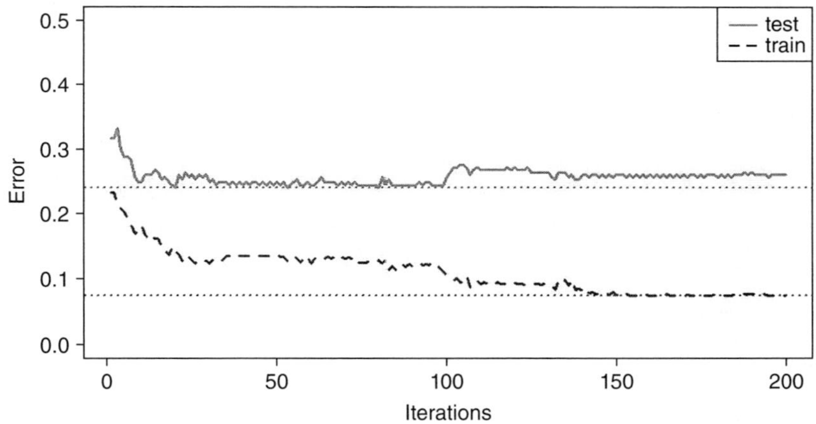

Figure 4.15 Error evolution in boosting.

in Figure 4.14, these are "checking account" (22.56%), "credit amount" (18.37%), "purpose" (11.79%), and "duration" (11.04%).

```
> importanceplot(german.boosting, horiz=FALSE, cex.names=.9)
```

The estimate can still be improved, especially in the test set, by pruning the combination. Observing the evolution of the error (Figure 4.15) we stop at iteration number 20, where the minimum is reached for this set, to improve the generalization capacity.

```
> # Error Evolution
> errorevol(german.boosting, german[train,])->errorevol.train
> errorevol(german.boosting, german[-train,])->errorevol.test
>
> plot.errorevol(errorevol.test, errorevol.train, main="Boosting error vs
number of trees")
> abline(h=min(errorevol.test[[1]]), col="red",lty=3)
> abline(h=min(errorevol.train[[1]]), col="blue",lty=3)
>
> newmfinal = which(errorevol.test[[1]]==min(errorevol.test[[1]]))
> newmfinal[1]

[1] 20
```

```
> errorevol.train$error[20]

[1] 0.1413333

> errorevol.test$error[20]

[1] 0.24
```

The minimum error in the test set is obtained for a combination of 20 trees, as we said before, and the errors obtained in the training and test sets for boosting pruned to 20 trees are 14.13% and 24%, which has improved substantially the results of both the individual tree and the pruned bagging combination. The confusion matrices for both sets are shown below.

```
> #Training set
> german.predboost<- predict.boosting(german.boosting, newdata=german[train,],
+ newmfinal=newmfinal[1])
> german.predboost$confusion

               Observed Class
Predicted Class Good Bad
           Bad    39 147
           Good  497  67

> german.predboost$error

[1] 0.1413333

> #Test set
> german.predboost<- predict.boosting(german.boosting, newdata=german[-train,],
+ newmfinal=newmfinal[1])
> german.predboost$confusion

               Observed Class
Predicted Class Good Bad
           Bad    17  43
           Good  147  43

> german.predboost$error

[1] 0.24
```

4.3.4 Combination using Random Forest

To apply the random forest algorithm, we use the `randomForest` R package (Liaw et al., 2002). The current version, 4.6-12, has been available since 2015. The description of the function says "`randomForest` implements Breiman's random forest algorithm (based on Breiman and Cutler's original Fortran code) for classification and regression. It can also be used in unsupervised mode for assessing proximities among data points".

Although it has more than 20 arguments, we focus on just two of them, the number of trees, `ntree`, and the number of variables randomly sampled as candidates at each split, `mtry`. For classification, the default number of drawn variables is set to the squared root of the total number of predictor variables.

We start building the model in the training set with 200 trees, as in the previous sections. The print function shows a summary of the random forest model. In this case, as there are 20 predictor variables, four variables are randomly selected as candidates at each split. The OOB estimation of the error rate is 23.1%, but differs a lot between the classes: 7.65% for the "Good" class and 61.7% for the "Bad" one. In the confusion matrix there are 41 Good clients classified as Bad and 132 Bad clients misclassified, leading to a total of 173 errors.

```
> german.rf <- randomForest(formula=formula, data=german[train,], ntree = 200,
+ importance =TRUE)
> print(german.rf)

Call:
 randomForest(formula = formula, data = german[train,], ntree = 200,
importance = TRUE)
               Type of random forest: classification
                     Number of trees: 200
No. of variables tried at each split: 4

        OOB estimate of  error rate: 23.07
Confusion matrix:
     Good Bad class.error
Good  495  41  0.07649254
Bad   132  82  0.61682243
```

Next, we present the results in the training and test sets. The training error is 0 and the test set is 25.6%. There are 7 Good and 57 Bad clients wrongly classified in the test set.

```
> # Prediction on the train set
> german.predrf <- predict(german.rf,newdata = german[train,],type="class")
> print(cm<-table(observed = german[train, "class"], predicted = german.predrf))

        predicted
observed Good Bad
    Good  536   0
    Bad     0 214

> print((cm[2,1]+cm[1,2])/sum(cm[1:2,1:2]))

[1] 0

> # Prediction on the test set
> german.predrf <- predict(german.rf,newdata = german[-train,],type="class")
> print(cm<-table(observed = german[-train, "class"], predicted = german.predrf))

        predicted
observed Good Bad
    Good  157   7
    Bad    57  29

> print((cm[2,1]+cm[1,2])/sum(cm[1:2,1:2]))

[1] 0.256

> importance(german.rf)

                          Good        Bad MeanDecreaseAccuracy
MeanDecreaseGini
checking.account    10.7297751б 16.7517690            18.7325249       31.4590241
duration             8.21480337  4.7920187             9.6551830       27.5609968
credit.history       5.66638692  5.1666633             7.6590539       20.2895733
purpose              4.92753038  4.2402104             6.4515544       26.7305599
credit.amount        7.39254848  2.5825193             7.2056709       39.9832395
savings.account      2.68621893  7.1986912             6.6193690       15.4114625
employment           1.95845893 -0.3253084             1.4700667       17.2601860
installment.rate     1.55944839 -1.0929275             0.5370280       11.6451116
personal.status      1.66676063  1.9485400             2.6436661       10.4136351
debtors              4.04593876  4.4194227             5.7799097        5.4655383
residence            1.16825108  0.5398511             1.2710136       12.8194446
property             3.43706471 -1.5740714             1.8621631       13.3284300
age                  2.79947736  2.8307622             3.8216888       28.7681359
```

```
other.installment.plans  3.30989788  3.1671250            4.4321688          8.8949319
housing                  0.92697943 -0.2640968            0.4357880          6.6841643
existing.credits         2.47789693  0.6478151            2.7124819          6.4411195
job                      2.57593223  0.1732419            2.2560587          9.1376091
maintenance.for          1.01609618  0.3592283            0.9716095          3.6518826
telephone                1.02086005  0.8410001            1.4635418          4.3613398
foreign.worker          -0.03657412  1.4640701            0.8975905          0.9017336
```

```
> barplot(german.rf.imp[,4][order(german.rf.imp[,4],decreasing=TRUE)],
+ main="Variables relative importace", + col="lightblue",las=1,xaxs="r")
```

The importance function returns a matrix where the first two columns are the class-specific measures computed as mean decrease in accuracy. The third and fourth columns are the mean decrease in accuracy and in the Gini index, respectively, over all classes. In all cases, "checking account" is the most important variable, but the second most important for classifying the Good clients is "duration" and for Bad clients is "saving accounts". Figure 4.16 shows these results.

The evolution of the OOB error can be observed using the plot function on the randomForest object. In this case the OOB estimations of the error by class are also shown (Figure 4.17). It can be seen that the OOB error for the class "Good" decreases as the forest grows and the opposite occurs with the class "Bad" (lines at the bottom and top of the figure, respectively).

The function `getTree` extracts the structure of a tree of the randomForest object in a dataframe. Here we select the first tree ($k = 1$) just as an example. There are 261 nodes, one on each row, but only nodes from 1 to 5 and from 250 to 261 are shown for simplicity. The information shows which are the left and right daughter nodes, the variable selected for the split and the exact point (value), the status indicates if it is a terminal node (−1) or not (1), and finally the class predicted if it is a terminal or leaf node. For leaf nodes, the first two columns and split point are 0 and the split variable is set as <NA>.

```
> tree1<-getTree(german.rf, k=1,labelVar=TRUE)
> dim(tree1)
```

```
[1] 261   6
```

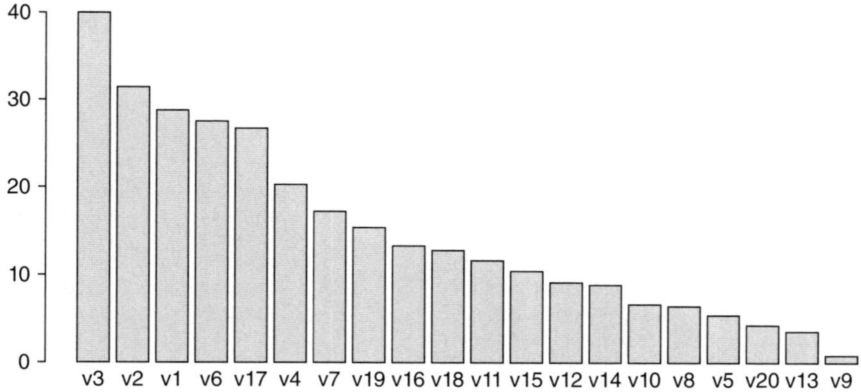

Figure 4.16 Variable relative importance in random forest.

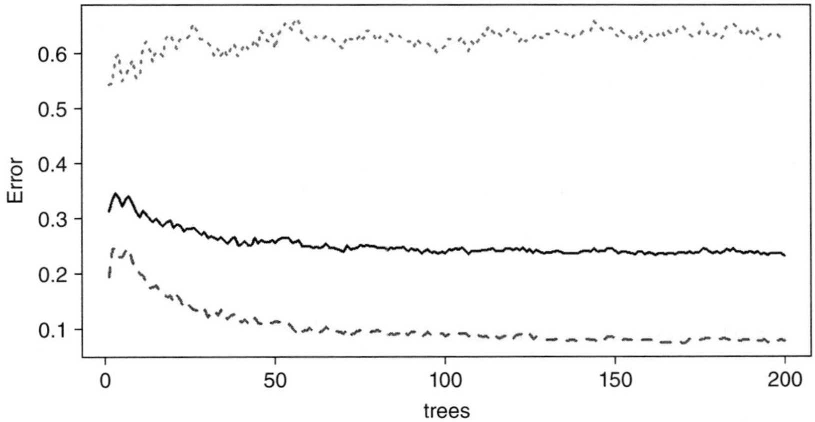

Figure 4.17 OOB error evolution in random forest.

```
> tree1[1:5,]

  left daughter right daughter          split var split point status prediction
1             2              3     credit.history          15      1       <NA>
2             4              5     credit.history           3      1       <NA>
3             6              7            purpose        1287      1       <NA>
4             8              9   checking.account           1      1       <NA>
5            10             11         employment           3      1       <NA>

> tree1[250:261,]

    left daughter right daughter          split var split point status prediction
250           258            259            purpose      1919.0      1       <NA>
251             0              0               <NA>         0.0     -1       Good
252             0              0               <NA>         0.0     -1        Bad
253             0              0               <NA>         0.0     -1       Good
254             0              0               <NA>         0.0     -1       Good
255           260            261   installment.rate         1.5      1       <NA>
256             0              0               <NA>         0.0     -1       Good
257             0              0               <NA>         0.0     -1        Bad
258             0              0               <NA>         0.0     -1        Bad
259             0              0               <NA>         0.0     -1       Good
260             0              0               <NA>         0.0     -1       Good
261             0              0               <NA>         0.0     -1        Bad
```

As has been seen using the default value for most of the arguments leads to trees larger than those previously used for bagging or boosting. Next, we train another random forest setting `maxnode=8`, which is equivalent to `maxdepth=3` used for bagging and boosting. The same information as above is shown below, and it should be highlighted that the test error increases to 33.2%.

```
> german.rf <- randomForest(formula=formula, data=german[train,], ntree = 200,
+ importance =TRUE, maxnodes=8)
> print(german.rf) # a summary of the output

Call:
 randomForest(formula = formula, data = german[train,], ntree = 200,
 importance = TRUE, maxnodes = 8)
               Type of random forest: classification
                     Number of trees: 200
No. of variables tried at each split: 4
```

```
        OOB estimate of error rate: 28
Confusion matrix:
     Good Bad class.error
Good  535   1 0.001865672
Bad   209   5 0.976635514
```

```
> # Prediction on the train set
> german.predrf <- predict(german.rf,newdata = german[train,],type="class")
> print(cm<-table(observed = german[train, "class"], predicted = german.predrf))
```

```
        predicted
observed Good Bad
    Good  536   0
    Bad   202  12
```

```
> print((cm[2,1]+cm[1,2])/sum(cm[1:2,1:2]))
```

```
[1] 0.2693333
```

```
> # Prediction on the test set
> german.predrf <- predict(german.rf,newdata = german[-train,],type="class")
> print(cm<-table(observed = german[-train, "class"], predicted = german.predrf))
```

```
        predicted
observed Good Bad
    Good  164   0
    Bad    83   3
```

```
> print((cm[2,1]+cm[1,2])/sum(cm[1:2,1:2]))
```

```
[1] 0.332
```

```
> importance (german.rf)
```

```
                                Good        Bad MeanDecreaseAccuracy MeanDe-
creaseGini
checking.account          3.47714273  7.5774287           6.92925203      15.98733643
duration                  6.09260017  3.9304441           6.08405101       4.98572284
credit.history            5.74830812  7.1665207           7.95683859       6.09425809
purpose                   2.22877367  2.9166244           3.63929994       4.29185675
credit.amount             5.69581252  3.4931493           6.13817028       5.94894934
savings.account           3.37671469  4.5791623           5.19543437       4.59978793
employment                0.97536755 -1.2091493           0.03611812       1.58039867
installment.rate          1.88403474 -1.9243444           1.01939269       0.48399317
personal.status           1.33574869  1.1033534           1.68107774       0.96953642
debtors                   1.53352886  2.8661684           2.38733810       0.79518565
residence                -0.81565151  1.2352173           0.22590149       0.39727847
property                  2.03287897  1.3717692           2.48964090       1.20995648
age                       3.30871353  3.3023955           4.33339118       3.14894373
other.installment.plans   2.35769384  1.2676152           2.64148200       1.44456545
housing                   1.55299091  1.5591571           2.10031104       1.26681133
existing.credits          0.47624904 -0.2650579           0.49084403       0.13735281
job                       1.12913318 -1.7090237          -0.61612566       0.58379519
maintenance.for          -0.18885088 -1.2671016          -0.87061237       0.08974366
telephone                -0.06553806 -0.4281075          -0.37521880       0.08999633
foreign.worker            0.46624153  1.0025094           0.81065528       0.10058016
```

```
> tree1<-getTree(german.rf, k=1,labelVar=TRUE)
> dim(tree1)
```

```
[1] 15  6
```

```
> tree1[1:15,]
```

```
  left daughter right daughter        split var split point status prediction
1             2              3 checking.account         3.0      1       <NA>
2             4              5          debtors         3.0      1       <NA>
3             6              7          purpose      1673.0      1       <NA>
4             8              9              age        24.5      1       <NA>
```

```
5           10          11  savings.account      2.0      1       <NA>
6           12          13    credit.amount   4441.5      1       <NA>
7           14          15          debtors      2.0      1       <NA>
8            0           0             <NA>      0.0     -1        Bad
9            0           0             <NA>      0.0     -1       Good
10           0           0             <NA>      0.0     -1        Bad
11           0           0             <NA>      0.0     -1       Good
12           0           0             <NA>      0.0     -1       Good
13           0           0             <NA>      0.0     -1       Good
14           0           0             <NA>      0.0     -1       Good
15           0           0             <NA>      0.0     -1       Good
```

4.3.5 Cross-Validation Comparison

The results obtained by any of the above methods, individual classification tree, combination by bagging, boosting or random forest, can vary considerably depending on the individuals included in the training and test sets. In each specific problem, the researcher chooses the best tree obtained or the best of the ensemble combinations or the best of any other classification method available. If we want to compare several methods in a specific problem and minimize their dependence on the individuals that compose the training and test sets, we should compare them by cross-validation.

We can compare bagging, boosting, and random forest by cross-validation by choosing the number of folders into which we want to divide the complete data set, usually 10. We present the comparison using the same parameters as in the three previous sections.

```
> german.baggingcv <- bagging.cv(formula=formula, v=10, data=german, mfinal=mfinal,
+ control=control)
> german.baggingcv[-1]

$confusion
               Observed Class
Predicted Class Good Bad
           Bad    57 101
           Good  643 199

$error
[1] 0.256

> german.boostcv<-boosting.cv(formula=formula, data=german, v=10, mfinal=mfinal,
boos=T,
+ control=control,coeflearn=coeflearn)
> german.boostcv[-1]

$confusion
               Observed Class
Predicted Class Good Bad
           Bad    93 149
           Good  607 151

$error
[1] 0.244

> #CV error rate
> german.rfcv <- rfcv(german[,1:20], german$class, cv.fold=10, maxnode=8)
> names(german.rfcv)

[1] "n.var"      "error.cv"   "predicted"

> print(cm<-table(observed = german[, "class"], predicted =
german.rfcv$predicted$'20'))
```

```
        predicted
observed Good Bad
    Good  694   6
    Bad   284  16

> print((cm[2,1]+cm[1,2])/sum(cm[1:2,1:2]))

[1] 0.29
```

We can see that, for this problem, boosting again overcomes bagging and random forest since the average error in cross-validation using 10 folders is 29% for random forest, 25.6% for bagging, and 24.4% for boosting, which means an improvement of approximately 5%.

5

Bankruptcy Prediction Through Ensemble Trees

Esteban Alfaro, Matías Gámez, and Noelia García

5.1 Introduction

Corporate failure prediction is traditionally an important management science task but became even more important during the last economic crisis. This is a typical classification problem where the objective is to determine which indicators are involved in the failure or success of a corporation. Ensemble methods have proved to be useful in addressing this question.

Despite the complexity of the matter, a two-class problem has usually been considered to tackle this classification problem. For this reason, in this chapter we will consider both the dichotomous and multiclass perspectives. Much of the information collected in this chapter is excerpts of works already published to which we refer for more detailed information: Multiclass Corporate Failure Prediction by Adaboost.M1 (Alfaro et al., 2007), Linear Discriminant Analysis versus AdaBoost for Failure Forecasting (Alfaro et al., 2008a). and Bankruptcy Forecasting: An Empirical Comparison of AdaBoost and Neural Networks (Alfaro et al., 2008b).

5.2 Problem Description

Corporate failure prediction consists of separating the firms with a high probability of future failure from those that are considered to be healthy. In order to do this, it is necessary to know the financial state of the firms. This is a matter that has been studied for almost four decades and is still one that concerns the scientific community. While there have been many methodological improvements during this time, these have not always led to an improvement in results. Nowadays, the subject is attracting the attention of many researchers due to the wider availability of financial information. Another factor that has contributed to the increasing interest is the use of alternative analysis tools from machine learning. Moreover, qualitative attributes are being tested that, when combined with traditional numerical ones, improve the accuracy of the forecasting models and knowledge of the factors which determine whether a firm will survive or not.

According to Gallego and Gómez (2002), two main research lines can be distinguished among the numerous studies into failure forecasting: on the one side, that which focuses on finding classification methods with better prediction accuracy, and

Ensemble Classification Methods with Applications in R, First Edition.
Edited by Esteban Alfaro, Matías Gámez, and Noelia García.

on the other, that which focuses on the most relevant ratios for minimizing prediction error. Unfortunately, experts do not agree on the best ratios to select to solve the problem. The first line encompasses this work since it proposes a novel method that is well known in the field of artificial intelligence due to its high levels of accuracy, with an improvement in the results provided by other more traditional methods.

Since Beaver's pioneering work (Beaver, 1966) that solved the failure prediction task using financial ratios in a univariate way, a great number of papers have attempted to improve on the results. Altman (1968) carried out pioneering work with the application of multivariate discriminant analysis. Even today, this model still constitutes the reference point for comparing the results of new contributions.

After this pioneering research, other models, such as conditional probability, logit (Ohlson, 1980), and probit (Zmijewski, 1984), were applied to avoid the very restrictive hypothesis regarding the normality and equality of variance-covariance matrices. More recently, classification trees (Frydman et al., 1985; Gabás Trigo, 1990) and artificial neural networks (Wilson and Sharda, 1994; Serrano-Cinca, 1997) have been proposed as alternative methods for failure prediction.

As mentioned previously, the comparison of different classification methods is a very important research line that constitutes the goal of many papers. On an international level, we can mention the work by Charalambous et al. (2000) that applied several versions of artificial neural networks to a set of 139 pairs of American failed and healthy firms for the period 1983–1994. More specifically, these authors compared the generalization ability of six methods, radial basis function, feed-forward nets using the conjugate gradient algorithm, the learning vector quantization (LVQ) method, the back-propagation algorithm, self-organizing feature maps (SOFMs) proposed by Kohonen, and logistic regression, with the feed-forward networks and SOFMs obtaining the best results with accuracies for the test data sets of 82.6% and 80.2%, respectively.

Other interesting publications include Pérez (2006), Kumar and Ravi (2007), and Tkáč and Verner (2016), which collected the studies relating to failure prediction using artificial neural networks in the first case and several statistical techniques in addition to methods from artificial intelligence in the second. More recent papers have dealt with corporate failure prediction in the presence of imbalanced data sets (Zhou, 2013; Zięba et al., 2016) or missing data (Zhou and Lai, 2017).

We should also mention the work by Serrano-Cinca (1997) that compared discriminant analysis and artificial neural networks by means of the leaving-one-out cross-validation method using a sample of 66 Spanish banks (29 failed and 37 healthy). By way of conclusion, the neural model performs better with an accuracy of 93.94%, 7.58 points above the 86.36% reached by discriminant analysis.

Ferrando Bolado and Blanco Ramos (1998) compared the discriminant and logit model abilities for failure prediction in the Comunidad Valenciana. As we have in our work, they used legal failure as surrogate failure, working with companies in receivership or bankrupt in the 1992–1994 period. The sample consisted of 88 failed firms and another 88 healthy ones. The total sample was divided into two sets: the training data set with 120 cases and the validation set with 56 cases, maintaining the parity between classes. The discriminant function obtained an accuracy of 87.5% over the validation data set. On the other hand, the logit model obtained an accuracy of 85.71% over the validation sample, taking into account the information one year before the failure process. Both models, however, obtain the same results if the information is taken for the two years prior to failure. The authors concluded that the logit model is slightly better than discriminant analysis based on the type I and type II errors and the percentage of correctly classified cases in the estimation samples. Nevertheless, Ferrando and Blanco did not study the statistical significance of the differences, which can be carried out using any resampling technique.

Rodríguez López (2001) provided additional empirical evidence of the prediction of failure for small and medium-sized non-financial firms in Galicia. The concept of failure used in this work is wider than the legal concept of bankrupt firms or those in receivership and includes those firms involved in judicial processes involving debt collection or unpaid bills of exchange figuring in registries elaborated by specialized companies, all of a high number and amount. The sample consisted of 120 firms, half of which had failed and the other half were healthy. The validation sample comprised 29 failed firms and 284 healthy ones. Discriminant and logit analysis were the techniques used. One model was estimated for each of the four years prior to the moment of failure and an overall one was estimated from the information as a whole. The total percentages of cases correctly classified by the discriminant model over the validation sample were (from closer to farther from the time of failure) 97.4%, 88.5%, 73.2%, 71.6% and 75.9% for the overall model. The results achieved by the logit model were 96.5%, 84.6%, 80.8%, 69.2% and 89.2%, respectively. The logit model performed better in the cases of the overall model and the three years before the failure model.

At a later stage, de Llano Monelos et al. (2016) compared the effectiveness of eight popular forecasting methods. Their results indicated that, whatever the methodology, reliable predictions can be made using four variables that contain information about profitability, financial structure, turnover, and cash flows. Momparler et al. (2016) applied the boosted classification tree methodology to predict failure in the banking sector and identified four key scorecard variables to bank financial distress (size, Interbank ratio, income from non-operating items and net loans to deposits).

5.3 Applications

In this section, we firstly consider the dichotomous case, the most common, where we discriminate among "healthy" and "failed" firms. Secondly, we show a three-class example separating two types of failure, thus we have "failure1", "failure2" and "healthy" firms.

5.3.1 The Dichotomous Case

The companies used in this study were selected from the SABI database of Bureau Van Dijk (BVD), one of Europe's leading publishers of electronic business information databases and provider of the Wharton Research Data Services. SABI covers all the companies whose accounts are placed on the Spanish Mercantile Registry. In the case of failed firms, firms which had failed (bankruptcy and temporary receivership) during the period 2000–2003 were selected, but with the additional requirement that full information be provided about all the variables at the moment of failure and from the previous year. It is usual in failure prediction studies to select failed firms from various years in order to collect a higher sample size. There were therefore firms that had failed in years 2000, 2001, 2002 and 2003 so the information on variables should be understood in relative terms with respect to the moment of failure t with the previous year being $t - 1$. For a detailed study of the optimal forecasting horizon in bankruptcy prediction models see du Jardin (2017).

Healthy firms, on the other hand, were randomly selected from active companies at the end of 2003 with full data for 2003 and 2002. In this case, a second requirement was added: any firm with constantly negative profits during the last three years would be rejected, the reason being that even though they were still active in December 2003, they would soon enter a state of failure if they kept making a loss.

Within these requirements, 590 firms were selected for each group (failed/healthy), providing 1180 observations for the total data set. Instead of pairing the failed/healthy firms by sector or size as several failure studies have done, these variables were used as predictors in the selection process: the sector as a qualitative variable with 10 categories using the National Classification of Economic Activities (NACE-93 digit-1 level), and the size using the natural logarithm of total assets as a proxy variable. The legal structure was also used as a categorical explanatory variable with three options: public corporations, limited corporations, and other corporations.

In addition, 13 accounting-based ratios were included in the initial data set. In failure prediction studies, financial ratios are usually selected on the basis of three criteria: they should be commonly used in failure prediction literature, the information needed to calculate these ratios should be available, and the researchers' own decisions based on their experience in previous studies or on the basis of the preliminary trials. This study followed the same criteria. Sixteen predictor variables were therefore used for each company with information from the year prior to the failure and these variables are listed in Table 5.1.

Once the explanatory variables had been selected, we carried out an exploratory analysis of the quantitative data. Tables 5.2 and 5.3 show some statistics, normality tests and ANOVA for each variable and the correlation matrix. From these results, it is worth noting that none of the variables follows a normal distribution and that there is a high degree of correlation among the variables.

Table 5.1 Description of variables (some of these ratios are explained in White et al. (2003)).

Variable	Description	Variable	Description
CA.TA	Current assets/total assets	L.TD	Liabilities/total debt
CA.CL	Current assets/current liabilities	C.TA	Cash/total assets
EBIT.TA	Earnings before interest and taxes/ total assets	C.CL	Cash/current liabilities
CF.TD	Cash flow/total debt	S.CAP	Sales/capital
NACE1	NACE code at one digit	EBIT.CAP	Earnings before taxes/capital
WC.TA	Working capital/total assets	lnTA	Logarithm of total assets
WC.S	Working capital/sales	S.CA	Sales/current assets
LE	Legal structure	S.TA	Sales/total assets

Table 5.2 Results obtained from descriptive analysis and ANOVA (SW test = Shapiro-Wilks test; KS test = Kolmogorov-Smirnov test).

	EBIT.TA	EBIT.CAP	CA.TA	C.TA	S.CAP	CA.CL	C.CL
Mean	1.5514	0.5328	0.5197	0.1474	8.1580	2.7673	0.7119
Median	0.0784	0.3855	0.5984	0.0404	0.7804	1.1235	0.1025
Std. dev.	54.8164	4.3662	5.7313	1.8576	63.1374	17.1474	18.0535
Asymmetry	31.6723	14.3114	−24.9542	20.3904	2.8566	−2.4655	−20.7833
Skewness	1072.4834	305.8308	846.8311	733.5410	216.9099	182.5282	604.3463
Minimum	−394.3403	−25.1707	−179.9724	−27.4740	−1106.5820	−294.5847	−521.8325
Maximum	1840.4110	102.4375	66.5612	56.0455	1225.0620	283.4821	121.3837
1st quartile	0.0194	0.1390	0.1790	0.0117	0.0468	0.8083	0.0200
3rd quartile	0.2778	0.6984	0.8683	0.1022	8.1562	2.0993	0.5633
SW test	0.0180	0.1812	0.0493	0.0563	0.2205	0.2405	0.1053
SW p value	0.0000	0.0000	0.0000	0.0000	0.0000	0.0000	0.0000
KS test	0.3743	0.3558	0.4992	0.4992	0.4418	0.5559	0.4559
KS p value	0.0000	0.0000	0.0000	0.0000	0.0000	0.0000	0.0000
F value	0.9811	0.0486	1.7694	2.7768	12.4850	10.3950	1.3542
Prob. (>F)	0.3221	0.8256	0.1837	0.0959	0.0004	0.0013	0.2448

	WC.S	L.TD	CF.TD	WC.TA	S.TA	S.CA	lnTA
Mean	2.0219	−0.3856	0.3728	0.0187	1.0233	1.0065	7.5469
Median	0.0670	0.6832	0.0624	0.0620	0.8442	1.4837	7.4184
Std. dev.	17.2610	10.0915	10.1956	4.9913	1.3517	44.4035	1.5616
Asymmetry	0.6569	24.3215	−27.2205	−24.2789	−1.5633	−33.6943	0.5617
Skewness	95.7141	757.4138	872.4553	782.4589	36.3493	1147.4768	1.9395
Minimum	−240.3293	−46.8859	−324.0337	−154.2119	−16.2484	−1514.6670	1.4081
Maximum	239.3921	310.0544	60.1981	49.0374	10.4067	66.2963	15.0009
1st quartile	−0.0786	−1.1317	0.0114	−0.0474	0.0577	0.4463	6.5744
3rd quartile	0.7276	0.9118	1.5821	0.2064	1.6190	2.5122	8.4297
SW test	0.2706	0.1711	0.1228	0.0599	0.7327	0.0275	0.9745
SW p value	0.0000	0.0000	0.0000	0.0000	0.0000	0.0000	0.0000
KS test	0.2465	0.2792	0.3651	0.3122	0.4738	0.4858	0.9970
KS p value	0.0000	0.0000	0.0000	0.0000	0.0000	0.0000	0.0000
F value	19.339	20.226	9.4274	0.1121	294.38	1.0849	0.0675
Prob. (>F)	0.0000	0.0000	0.0022	0.7378	0.0000	0.2978	0.7951

In this section, the same failure prediction problem is solved using three different classification methods to compare their classification accuracies in this task. To estimate the real accuracy, the total initial sample of 1180 Spanish companies was divided into two sets: 80% were used as a training set to build the classifier, and the rest were hidden from the classification method and were presented as new data to check the prediction accuracy. We therefore had 1180 companies, 80% of which (944) were included in the

Table 5.3 Correlation matrix.

	EBIT.TA	EBIT.CAP	CA.TA	C.TA	S.CAP	CA.CL	C.CL
EBIT.TA	1.0000	0.0356	0.5241	0.9500	−0.0035	0.0019	0.0031
EBIT.CAP	0.0356	1.0000	0.0122	0.0309	0.2833	0.0011	0.0019
CA.TA	0.5241	0.0122	1.0000	0.6997	0.0067	0.0084	−0.0007
C.TA	0.9500	0.0309	0.6997	1.0000	−0.0061	−0.0035	−0.0368
S.CAP	−0.0035	0.2833	0.0067	−0.0061	1.0000	−0.0119	−0.0044
CA.CL	0.0019	0.0011	0.0084	−0.0035	−0.0119	1.0000	0.2627
C.CL	0.0031	0.0019	−0.0007	−0.0368	−0.0044	0.2627	1.0000
WC.S	−0.0092	0.0115	0.0124	−0.0290	−0.0152	0.1035	0.0222
L.TD	−0.0121	0.0007	0.0015	−0.0252	0.0160	−0.0589	0.0936
CF.TD	0.1012	0.0105	0.0290	0.0950	−0.0106	0.0357	−0.0442
WC.TA	0.4728	0.0119	0.9603	0.6282	−0.0008	0.0197	0.0015
S.TA	−0.2322	0.0064	0.2638	−0.1172	0.1428	−0.0684	−0.0229

	WC.S	L.TD	CF.TD	WC.TA	S.TA	S.CA	lnTA
EBIT.TA	−0.0092	−0.0121	0.1012	0.4728	−0.2322	−0.0009	−0.0972
EBIT.CAP	0.0115	0.0007	0.0105	0.0119	0.0064	−0.0025	−0.0019
CA.TA	0.0124	0.0015	0.0290	0.9603	0.2638	0.0020	0.0107
C.TA	−0.0290	−0.0252	0.0950	0.6282	−0.1172	0.0015	−0.0936
S.CAP	−0.0152	0.0160	−0.0106	−0.0008	0.1428	0.0056	0.0366
CA.CL	0.1035	−0.0589	0.0357	0.0197	−0.0684	−0.0010	−0.0223
C.CL	0.0222	0.0936	−0.0442	0.0015	−0.0229	0.0019	0.0131
WC.S	1.0000	−0.0669	0.0612	0.0339	−0.0879	0.0062	−0.0104
L.TD	−0.0669	1.0000	−0.9526	0.0025	0.0993	0.0625	−0.0505
CF.TD	0.0612	−0.9526	1.0000	0.0322	−0.1002	−0.0552	0.0183
WC.TA	0.0339	0.0025	0.0322	1.0000	0.2098	−0.0019	0.0369
S.TA	−0.0879	0.0993	−0.1002	0.2098	1.0000	0.0435	−0.1790
S.CA	0.0062	0.0625	−0.0552	−0.0019	0.0435	1.0000	−0.0401
lnTA	−0.0104	−0.0505	0.0183	0.0369	−0.1790	−0.0401	1.0000

training set and the remaining 20% (236) formed the test set, with an equal number of healthy and failed firms (50% of each class) in both sets. As mentioned before, 16 diagnostic variables were used for each company with information for the year prior to the moment of failure.

We first built the model with linear discriminant analysis (LDA) using the stepwise method and the Wilks' lambda statistic as its criteria (see Alfaro et al. (2008a) for a more detail description of this process). Fisher (1936) proposed this method and a detailed explanation can be found in Uriel (1995). In order to apply the model, we used the statistical package SPSS 13.0 for Windows and the objective variable was coded with the values 0 for the failed firms and 1 for the healthy ones. We also included the qualitative variables NACE1 (with 10 categories) and LEGAL STRUCTURE (with three categories)

Table 5.4 Unstandardized coefficients of the canonical discriminant function.

Variable	D. Function 1
EBIT.TA	0.010
CA.TA	−0.097
L.TD	−0.016
S.TA	0.983
LnTA	0.164
NACE:−5	−0.566
PUBLIC CORP.	0.286
(Constant)	−2.168

as dummy variables creating the same number of variables as the number of categories minus one for each attribute (nine variables for NACE1 and two for LEGAL STRUCTURE). In each dummy variable, we assign the value 1 to those firms that show the corresponding modality and 0 to the remaining firms.

The selected variables for the final model were profitability, weight of the current asset, indebtedness, efficiency, size of the firm, sector with code 5 in NACE (wholesale and retail trade; repair of motor vehicles, motorcycles and personal/household goods), and whether the firm is a public corporation or not. Table 5.4 shows the unstandardized discriminant coefficients. These coefficients are used to assign a class to each observation. The firms that were most likely to fail were those with a high weight of current assets over total assets, those with a high ratio of indebtedness, and those which belong to NACE Sector 5. However, the firms that were most likely to remain active were those with higher profitability, those with a high efficiency ratio, larger firms, and those with the legal structure of a public corporation.

The Wilks' lambda statistic was computed to confirm that these variables have a significant joint capacity to discriminate between the two classes, yielding a value of 0.723. The p value in the chi-squared distribution associated with this statistic is 0.000, which allows us to reject the null hypothesis of equality of mean vectors between the two classes.

Table 5.5 shows the confusion matrix for both data sets and the corresponding errors. In this case, both errors are close to 20%: 20.763% in the training set and 20.339% in the test set. These error percentages are acceptable taking into account that we did not pair according to sector or size and there was therefore a certain degree of heterogeneity in the sample, which did, however, provide a wider level of applicability. The model therefore has an accuracy of 79.66% when it comes to predicting the future of new firms for the coming year. It is important to stress that for both sets the type I error is higher than the type II error.

The neural network model selected in this work is the well-known multilayer perceptron (Haykin, 1994; Bishop, 1995) with one hidden layer. This model was implemented using the TRAJAN software. The number of nodes in the input and the output layers was set by the structure of our analysis, i.e. the number of explanatory and output variables, respectively. On the other hand, several experiments were conducted to find the number of hidden elements, which gave the greatest accuracy in predicting the test

Table 5.5 Confusion matrix and errors with LDA.

Predicted	Real class			
class	Training	20.763%	Test	20.339%
LDA	Failed	Healthy	Failed	Healthy
Failed	357	81	84	14
Healthy	115	391	34	104

data set (see Alfaro et al. (2008b) for more details of this process). The results obtained from 20 different topologies, varying the number of hidden elements from 1 to 20 and training 10 times each topology, led to the selection of a 16:27-12-1:1 network, i.e. an input layer with 16 nodes, pre-processed into 27 nodes (the qualitative variables LE and NACE1 were encoded using the one-of-N method, which uses a set of dichotomous variables, one for each possible nominal value: three nodes were necessary for LE and 10 for NACE1 so the total number of input variables changed from 16 to 27), a hidden layer with 12 elements, and an output layer with one.

In order to train the network, the following decisions were set. First, the above mentioned division of the sample cases into two subsets: training and test. The activation functions were selected to be linear in the input layer and sigmoid in the hidden and output layers. The error function selected was the sum of squared errors and the learning algorithm to minimize it was back propagation with adaptive learning starting at 0.3 and finishing at 0.01 and momentum term set at the value 0.3. After 1868 epochs the network was trained and the following results were obtained. The main result from the confusion matrix in Table 5.6 is the error percentage, measured as the percentage of wrongly classified cases. The error was close to 11% for the training set and it was a little higher for the test set (12.712%). It is worth stressing that if we consider the most important error as the classification of a failed firm as a healthy one (type I error), the behaviour of the network was very satisfactory since it is about 4.025% in the training set (7.627% in the test set).

The relative contribution of each input variable to the global performance can be assessed by means of a sensitivity analysis, which entails testing the performance of the network as if the input variables were unavailable. This ratio measures the relation of the error if the correspondent input variable is unavailable with the error if all the variables

Table 5.6 Confusion matrix and errors with an artificial neural network.

Predicted	Real class			
class	Training	10.805%	Test	12.712%
MLP-ANN	Failed	Healthy	Failed	Healthy
Failed	453	83	109	21
Healthy	19	389	9	97

Table 5.7 Sensitive analysis.

Ranking	Variable	Ratio
1	S.TA	1.3091
2	NACE1	1.2109
3	LnTA	1.1585
4	LE	1.1311
5	L.TD	1.0964

are available. A ratio of one or lower therefore means that pruning this input variable has no effect on the performance of the network. Table 5.7 shows the ranking of the input variables with a ratio above 1.05, which means that these variables can be considered as the most important ones in terms of their discriminatory power. According to the sensitive analysis results, the most important ratios are the efficiency (S.TA), the sector, the proxy variable for the firm size, the legal structure, and the level of indebtedness (L.TD).

Using the functions described in Chapter 4 an AdaBoost classifier was built with 100 trees pruned using a maximum depth of 2 to limit the size of the individual tree in each AdaBoost epoch. The depth of the tree is the distance between a leaf node and the root node. The R code to train the AdaBoost model and its output is shown below.

```
> sabi.boosting <- boosting(Class ~ ., data=sabi[ind,], mfinal=100, maxdepth=2)
> sabi.boosting

$formula
ESTADO ~ .


$trees
$trees[[1]]
n= 944

node), split, n, loss, yval, (yprob)
      * denotes terminal node

1) root 944 451 Healthy (0.47775424 0.52224576)
  2) L.TD>=0.3238577 583 136 Failed (0.76672384 0.23327616)
    4) CF.TD>=-2.052969 552 106 Failed (0.80797101 0.19202899) *
    5) CF.TD< -2.052969 31   1 Healthy (0.03225806 0.96774194) *
  3) L.TD< 0.3238577 361   4 Healthy (0.01108033 0.98891967)
    6) CF.TD< 0.02723354 4   0 Failed (1.00000000 0.00000000) *
    7) CF.TD>=0.02723354 357   0 Healthy (0.00000000 1.00000000) *

$trees[[2]]
n= 944

node), split, n, loss, yval, (yprob)
      * denotes terminal node

1) root 944 376 Healthy (0.398305085 0.601694915)
  2) L.TD>=0.3379687 628 253 Failed (0.597133758 0.402866242)
    4) C.CL< 0.1289769 424 119 Failed (0.719339623 0.280660377) *
    5) C.CL>=0.1289769 204  70 Healthy (0.343137255 0.656862745) *
  3) L.TD< 0.3379687 316   1 Healthy (0.003164557 0.996835443) *

...
```

```
 $trees[[99]]
n= 944

node), split, n, loss, yval, (yprob)
      * denotes terminal node

1) root 944 435 Healthy (0.4608051 0.5391949)
  2) L.TD>=0.7777208 510 221 Failed (0.5666667 0.4333333)
    4) NACE1=1,3,4,5,6,7,9 370 127 Failed (0.6567568 0.3432432) *
    5) NACE1=0,2 140  46 Healthy (0.3285714 0.6714286) *
  3) L.TD< 0.7777208 434 146 Healthy (0.3364055 0.6635945)
    6) lnTA>=7.43167 132  60 Failed (0.5454545 0.4545455) *
    7) lnTA< 7.43167 302  74 Healthy (0.2450331 0.7549669) *

$trees[[100]]
n= 944

node), split, n, loss, yval, (yprob)
      * denotes terminal node

1) root 944 385 Healthy (0.40783898 0.59216102)
  2) EBIT.TA< -0.1085696 68  10 Failed (0.85294118 0.14705882)
    4) NACE1=0,2,4,5,6,7,8,9 59   4 Failed (0.93220339 0.06779661) *
    5) NACE1=1 9   3 Healthy (0.33333333 0.66666667) *
  3) EBIT.TA>=-0.1085696 876 327 Healthy (0.37328767 0.62671233) *

$weights
[1] 1.06649115 0.87397842 0.92619205 0.56640183 0.36014186 0.31979554 0.80218088
  [8] 0.63803383 0.32448798 0.55781699 0.06151817 0.71257360 0.19964476 0.23741312
 [15] 0.51050552 0.06577265 0.00100000 1.03378950 0.94916938 0.84119346 0.29884387
 [22] 0.89526207 0.00100000 1.01287652 0.84521105 0.86979741 0.64739732 0.46000066
 [29] 0.33864949 0.56067027 0.44194365 0.51596046 1.01287652 0.06577265 0.22178670
 [36] 0.00100000 0.99261363 0.89959722 0.90834921 0.36979634 0.61353426 0.14940696
 [43] 0.70246836 0.29884387 0.29884387 0.43938627 0.47570414 0.54648526 1.04993134
 [50] 0.00100000 0.97295507 0.89526207 0.85331360 0.44194365 0.63184602 0.49162910
 [57] 0.45481325 0.70246836 0.29191705 1.02324869 0.15808349 0.15591219 0.14724136
 [64] 0.03815410 0.13643349 0.03603254 0.03603254 0.30115898 0.76872327 0.15157395
 [71] 0.12781007 0.24189848 0.00100000 0.97781538 0.81754017 0.88667137 0.94916938
 [78] 0.16896260 0.14724136 0.62876812 0.15591219 0.28501736 0.57506291 0.00100000
 [85] 1.01804192 1.07777594 0.35773833 0.61353426 0.85739921 0.28043224 0.60152748
 [92] 0.53806972 0.15374235 0.43428763 0.08068310 0.81754017 0.17114310 0.20847720
 [99] 0.16243054 0.13859245
$votes
         [,1]      [,2]
[1,] 39.34045  9.80669
[2,] 35.33347 13.81367
[3,] 38.27035 10.87679
[4,] 36.64356 12.50358
[5,] 37.93854 11.20860
...

[475,] 21.4964030 27.65073
[476,]  0.5933996 48.55374
[477,]  0.0020000 49.14514
[478,]  0.2025846 48.94455
[479,] 15.7610410 33.38610
[480,]  7.2334756 41.91366
...

[940,]  4.2315235 44.91561
[941,] 18.6462053 30.50093
[942,]  0.6963263 48.45081
```

```
[943,]  0.2025846 48.94455
[944,] 32.4256978 16.72144

$prob
          [,1]      [,2]
[1,] 0.8004626 0.1995374
[2,] 0.7189324 0.2810676
[3,] 0.7786892 0.2213108
[4,] 0.7455889 0.2544111
[5,] 0.7719380 0.2280620
...
              [,1]      [,2]
[475,] 4.373887e-01 0.5626113
[476,] 1.207394e-02 0.9879261
[477,] 4.069413e-05 0.9999593
[478,] 4.122003e-03 0.9958780
[479,] 3.206909e-01 0.6793091
[480,] 1.471800e-01 0.8528200
...
              [,1]      [,2]
[940,] 0.086099083 0.9139009
[941,] 0.379395549 0.6206045
[942,] 0.014168197 0.9858318
[943,] 0.004122003 0.9958780
[944,] 0.659767777 0.3402322

$class
  [1] "Failed" "Failed" "Failed" "Failed" "Failed" "Failed" "Failed" "Failed"
  [9] "Failed" "Failed" "Healthy" "Failed" "Failed" "Failed" "Failed" "Failed"
 [17] "Failed" "Failed" "Failed" "Failed" "Failed" "Failed" "Failed" "Failed"
...
[465] "Failed" "Failed" "Failed" "Failed" "Failed" "Failed" "Failed" "Failed"
[473] "Healthy" "Healthy" "Healthy" "Healthy" "Healthy" "Healthy" "Healthy" "Healthy"
[481] "Healthy" "Healthy" "Healthy" "Healthy" "Failed" "Healthy" "Failed" "Failed"
[489] "Healthy" "Healthy" "Healthy" "Healthy" "Healthy" "Healthy" "Failed" "Healthy"
...
[921] "Healthy" "Healthy" "Healthy" "Healthy" "Healthy" "Healthy" "Healthy" "Healthy"
[929] "Healthy" "Healthy" "Healthy" "Healthy" "Healthy" "Healthy" "Healthy" "Healthy"
[937] "Healthy" "Healthy" "Healthy" "Healthy" "Healthy" "Healthy" "Healthy" "Failed"

$importance
 LE       NACE1    EBIT.TA   EBIT.CAP    CA.TA     C.TA      S.CAP     CA.CL     C.CL
 2.888087 8.303249 17.328520 3.610108 5.054152 5.054152 2.166065 1.805054 8.664260
 WC.S     L.TD     CF.TD     WC.TA      S.TA      S.CA      lnTA
 1.444043 13.718412  6.498195  2.166065 10.469314  1.805054  9.025271
```

The *sabi.boosting* object can be used to predict the class of the sabi examples belonging to the test set, which also gives the confusion matrix and the test error. In the test set, four failed firms were classified as healthy and 17 healthy firms were misclassified, so the test error reached 8.9%.

```
> sabi.predboosting<- predict.boosting(sabi.boosting, newdata=sabi[-ind,])
> sabi.predboosting

$formula
Class ~ .

$votes
             [,1]       [,2]
  [1,] 39.84507509  9.3020627
  [2,] 27.14754813 21.9995897
  [3,] 34.60618245 14.5409554
```

```
  [4,] 26.35330460 22.7938332
  [5,] 25.46903257 23.6781052
  ...
[115,] 41.31470024  7.8324376
[116,] 45.58482314  3.5623147
[117,] 32.66988494 16.4772529
[118,] 35.63263005 13.5145078
[119,]  0.00200000 49.1451378
[120,]  6.25363841 42.8934994
...
[230,]  3.89247237 45.2546654
[231,]  0.24376446 48.9033734
[232,]  0.59339959 48.5537382
[233,]  4.92346720 44.2236706
[234,]  0.86746944 48.2796684
[235,]  2.42698972 46.7201481
[236,]  0.85801048 48.2891273

$prob
               [,1]       [,2]
  [1,] 8.107303e-01 0.18926967
  [2,] 5.523729e-01 0.44762708
  [3,] 7.041342e-01 0.29586576
  [4,] 5.362124e-01 0.46378760
  [5,] 5.182201e-01 0.48177994
  ...
[115,] 8.406329e-01 0.15936712
[116,] 9.275174e-01 0.07248265
[117,] 6.647363e-01 0.33526373
[118,] 7.250194e-01 0.27498057
[119,] 4.069413e-05 0.99995931
[120,] 1.272432e-01 0.87275681
...
[230,] 7.920039e-02 0.92079961
[231,] 4.959891e-03 0.99504011
[232,] 1.207394e-02 0.98792606
[233,] 1.001781e-01 0.89982189
[234,] 1.765046e-02 0.98234954
[235,] 4.938212e-02 0.95061788
[236,] 1.745799e-02 0.98254201

$class
  [1] "Failed" "Failed" "Failed" "Failed" "Failed" "Failed" "Failed" "Failed"
  [9] "Failed" "Failed" "Failed" "Failed" "Failed" "Failed" "Failed" "Failed"
 [17] "Failed" "Failed" "Failed" "Failed" "Failed" "Failed" "Failed" "Failed"
 [25] "Failed" "Failed" "Failed" "Failed" "Failed" "Failed" "Failed" "Failed"
 [33] "Failed" "Failed" "Failed" "Failed" "Failed" "Healthy"  "Failed" "Failed"
 [41] "Failed" "Healthy"  "Failed" "Failed" "Failed" "Failed" "Failed" "Failed"
 [49] "Failed" "Failed" "Failed" "Failed" "Failed" "Failed" "Failed" "Failed"
 [57] "Failed" "Failed" "Failed" "Failed" "Failed" "Failed" "Failed" "Failed"
 [65] "Failed" "Failed" "Failed" "Failed" "Failed" "Failed" "Failed" "Failed"
 [73] "Failed" "Failed" "Failed" "Failed" "Failed" "Failed" "Failed" "Failed"
 [81] "Failed" "Failed" "Failed" "Failed" "Failed" "Failed" "Failed" "Failed"
 [89] "Failed" "Failed" "Failed" "Failed" "Failed" "Failed" "Failed" "Failed"
 [97] "Failed" "Failed" "Failed" "Failed" "Failed" "Failed" "Failed" "Failed"
[105] "Failed" "Failed" "Failed" "Failed" "Failed" "Failed" "Healthy"  "Failed"
[113] "Failed" "Healthy"  "Failed" "Failed" "Failed" "Failed" "Healthy"  "Healthy"
[121] "Healthy" "Healthy" "Healthy" "Healthy" "Healthy" "Healthy" "Healthy" "Healthy"
[129] "Healthy" "Failed" "Failed" "Healthy" "Healthy" "Failed" "Healthy" "Healthy"
[137] "Healthy" "Healthy" "Healthy" "Healthy" "Healthy" "Healthy" "Failed" "Healthy"
[145] "Healthy" "Failed" "Healthy" "Healthy" "Healthy" "Healthy" "Healthy" "Healthy"
[153] "Failed" "Healthy" "Healthy" "Healthy" "Healthy" "Healthy" "Healthy" "Healthy"
```

```
[161] "Healthy" "Healthy" "Healthy" "Healthy" "Healthy" "Healthy" "Healthy" "Healthy"
[169] "Healthy" "Failed" "Healthy" "Healthy" "Healthy" "Healthy" "Healthy" "Healthy"
[177] "Healthy" "Healthy" "Healthy" "Failed" "Healthy" "Failed" "Failed" "Healthy"
[185] "Healthy" "Failed" "Healthy" "Failed" "Healthy" "Failed" "Healthy" "Healthy"
[193] "Healthy" "Failed" "Healthy" "Failed" "Healthy" "Healthy" "Healthy" "Healthy"
[201] "Healthy" "Healthy" "Healthy" "Healthy" "Healthy" "Healthy" "Healthy" "Healthy"
[209] "Healthy" "Healthy" "Healthy" "Healthy" "Healthy" "Healthy" "Healthy" "Healthy"
[217] "Healthy" "Healthy" "Healthy" "Failed" "Healthy" "Healthy" "Healthy" "Healthy"
[225] "Healthy" "Failed" "Healthy" "Healthy" "Healthy" "Healthy" "Healthy" "Healthy"
[233] "Healthy"  "Healthy"  "Healthy"  "Healthy"

$confusion
               Observed Class
Predicted Class Failed Healthy
        Healthy      4     101
        Failed     114      17

$error
[1] 0.08898305
```

Table 5.8 shows the errors of the AdaBoost classifier in the training and test sets and their confusion matrices. This model achieved a training error of 7.63% and a test error of 8.898%. In addition, analysing the confusion matrix it can be seen that most of these errors were made because a healthy firm was classified as failed while the type I error was 3.39% in the test set.

The `boosting` function of the `adabag` package allows us to quantify the relative importance of the predictor variables. This is important because it is difficult to interpret the 100 trees used in the AdaBoost ensemble. This measure takes into account the gain of the Gini index given by a variable in each tree and its weight. Figure 5.1 shows all variables arranged from the greatest to least relative importance. In this case, the most important ratios are EBIT.TA, L.TD, S.TA and lnTA with values at this measure of 17.33, 13.72, 10.47 and 9.02%, respectively. On the other hand, the least important ratios are CA.CL, S.CA and WC.S with values below 2%.

Figure 5.2 shows the margin cumulative distribution for the AdaBoost classifier developed in this application. In this case, 7.63% and 8.89% of the negative margins matched the training error and test error, respectively. It should also be pointed out that about 25% of the observations had margins which were close to the unit (which shows those firms that had been classified with the highest degree of certainty).

Table 5.9 shows the results of four classification methods: LDA, N.Networks, AdaBoost and a single tree pruned by the rule "minimum error in cross-validation plus one standard deviation". Since the test error of AdaBoost was 8.898%, there was a

Table 5.8 Confusion matrix and errors with AdaBoost.

Predicted	**Real class**			
class	**Training**	**7.627%**	**Test**	**8,898%**
AdaBoost	**Failed**	**Healthy**	**Failed**	**Healthy**
Failed	460	60	114	17
Healthy	12	412	4	101

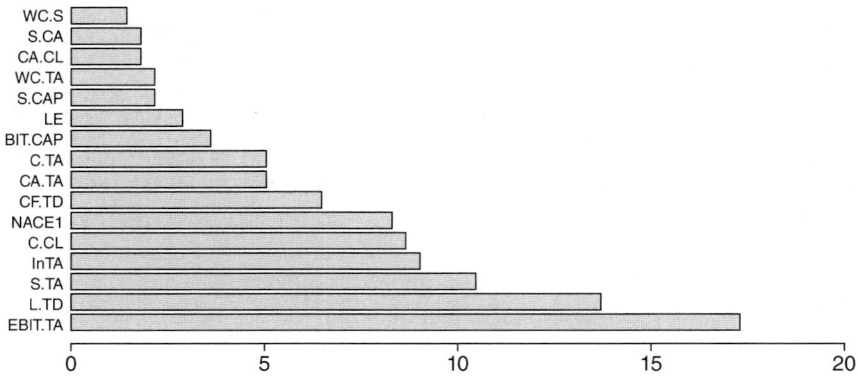

Figure 5.1 Variable relative importance in AdaBoost.

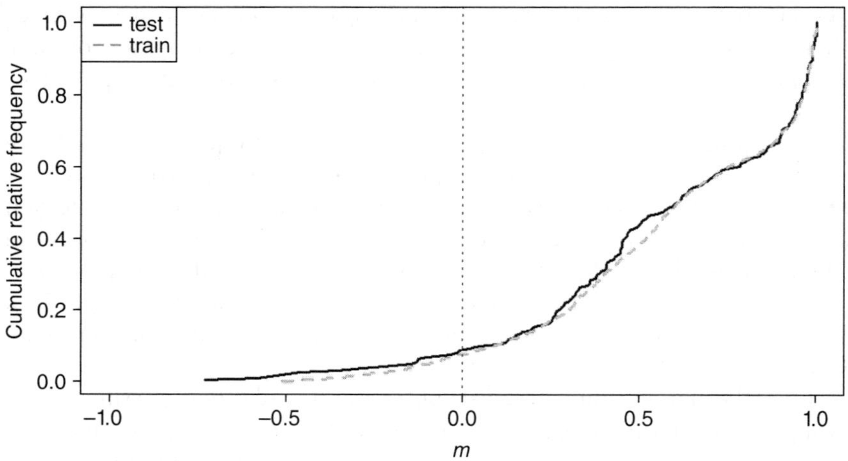

Figure 5.2 Margin cumulative distribution in AdaBoost.

Table 5.9 Comparison of results with other methods.

	Error rates					
	Training			Testing		
Models	**Overall**	**Type I**	**Type II**	**Overall**	**Type I**	**Type II**
LDA	20.763%	24.364%	17.161%	20.339%	28.814%	11.864%
C.Trees	10.452%	0.753%	20.155%	12.714%	3.390%	22.034%
ANN	10.805%	4.025%	17.585%	12.712%	7.627%	17.797%
AdaBoost	7.627%	2.523%	12.712%	8.898%	3.390%	14.407%

reduction of 30% compared with the neural network and the tree test error (12.712% and 12.714%) and a reduction of 56.25% in comparison to the LDA test error (20.34%).

5.3.2 The Three-Class Case

Although in corporate failure prediction literature only a binary classification problem is usually considered to differentiate failed from healthy firms, it makes sense to think that there is more than one type of failure. For instance, in the Spanish case only bankruptcy and temporary receivership firms are traditionally considered as failed firms, but it seems that there are other types of failure. In this section acquired and dissolved firms are also included as failed firms (Alfaro et al., 2007). We analyzed whether these firms have different financial behavior with respect to healthy firms. With this aim, the application worked with a three-class problem (healthy, failure1, and failure2), where failure1 includes acquired and dissolved firms, and failure2 includes bankruptcy and temporary receivership. The question here was how multiclass failure prediction problems can be faced in order to discriminate between two different states of failure in addition to the healthy class.

The companies in the sample were selected from the same database (SABI), during the same period, and with the same requirements as in the previous section. As in all studies on business failure, when selecting the sample attention is first focused on the less abundant class of companies, in this case category failure2, corresponding to companies that were bankrupt or in temporary receivership, of which there were only 590 companies in the 2000–2003 period that had complete information on all the variables to be used, both in the year of failure and in the previous five years.

In the class failure1 (companies acquired or dissolved) in the period 2000–2003 a total of 1014 companies were selected that had complete information on all variables, both for the year of failure and for the previous five years. Finally, the active companies are those that continued their activity in 2003 and they had information on all the variables for that year and the five previous years, so a total of 16,768 healthy companies was collected.

Once the information had been obtained, the outlier detection task was approached using the Mahalanobis distance from the observations to the center of their respective classes. For this calculation, the qualitative variables (NACE1 and legal structure) were left out. In this way, for each class, the observations that exceeded a certain threshold in the value of their distance from Mahalanobis were eliminated. This eliminated 161 companies in class failure1, 1499 active companies and 78 companies in class failure2.

In order to reduce the presence of noise, those healthy companies that had presented negative results in the last three years were eliminated because it was thought that although they were active companies, if they suffered continued losses they would soon fall into one of the failure modes, and therefore it would be difficult to use the classification method. For this reason, 95 active companies were eliminated, leaving the sample consisting of 853 failure1 companies, 15174 active companies and 512 failure2 companies.

As equal sizes were used for all classes, the restriction was imposed by class failure2, which had 512 observations, therefore samples of that size had to be extracted randomly for the healthy companies and the failure1 companies. The selection process did not pair firms of the three classes by sector and size because these variables are used as predictors. Consequently, in this case the total number of companies in the sample was 1536 (512 firms for each group failure1/failure2/healthy).

In this case, 18 predictor variables are used for each company with information for the year previous to the moment of failure. These variables are the same as used in section 5.3.1 (Table 5.1), adding the logarithm of current assets (lnCA) and net incomes/total assets (NI.TA).

The total initial sample consisted of 1536 Spanish companies, of which 90% were used as the training set and the rest as the test set. Therefore, the training set has 1383 observations and the test set has 153, keeping the same proportion as the classes. In the training set, a classifier was built both by the classification tree and the boosting method (AdaBoost.M1). Afterwards, these classifiers were tested in the remaining examples.

First, the tree is completely developed and then pruned using the 1-SE rule. As was explained in Chapter 4, this rule selects the smallest tree with a cross-validation error equal to or less than the minimum error plus a standard deviation. The R code and the results are shown below.

```
> sabi.rpart<-rpart(ESTADO~ .,method="class", data=sabi[ind,], cp=0,
minsplit=2, maxdepth=30)
> sabi.prune<-prune(sabi.rpart,cp=0.03)

> printcp(sabi.prune)

Classification tree:
rpart(formula = ESTADO ~ ., data = sabi[ind, ], method = "class", cp = 0, min-
split = 1,
maxdepth = 30)
Variables actually used in tree construction:
[1] lnTA L.TD
Root node error: 922/1383 = 0.66667
n= 1383
CP nsplit rel error xerror xstd
1 0.374187 0 1.00000 1.05748 0.018395
2 0.127983 1 0.62581 0.67245 0.020059
3 0.036876 2 0.49783 0.51302 0.019134
4 0.030000 3 0.46095 0.50217 0.019035
```

In this case, the pruned tree only used two variables. The first one was the indebtedness ratio (L.TD), selected in two splits as the best discriminant variable. The second was the proxy variable for the firm size (lnTA). At each node, a test is set, examples which satisfy the test being assigned to the left branch, while the rest go to the right. Then a label is assigned to each leaf (healthy, failure1, or failure2) and the number of failure1, failure2, and healthy firms present on it is shown. For instance, in the terminal leaf number 4 the class is failure1, and there are 185 failure1 firms, 59 healthy firms, and 67 failure2 firms. The tree structure was as follows and can be graphically seen in Figure 5.3.

```
> sabi.prune

n= 1383
node), split, n, loss, yval, (yprob)
* denotes terminal node
1) root 1383 922 Failure1 (0.33333333 0.33333333 0.33333333)
2) L.TD>=-0.01480628 1038 577 Failure1 (0.44412331 0.11175337 0.44412331)
4) L.TD < 0.6822301 311 126 Failure1 (0.59485531 0.18971061 0.21543408) *
5) L.TD >=0.6822301 727 333 Failure2 (0.37964237 0.07840440 0.54195323)
10) lnTA< 6.546879 188 89 Failure1 (0.52659574 0.12765957 0.34574468) *
11) lnTA>=6.546879 539 210 Failure2 (0.32838590 0.06122449 0.61038961) *
3) L.TD < -0.01480628 345 0 Healthy (0.00000000 1.00000000 0.00000000) *
```

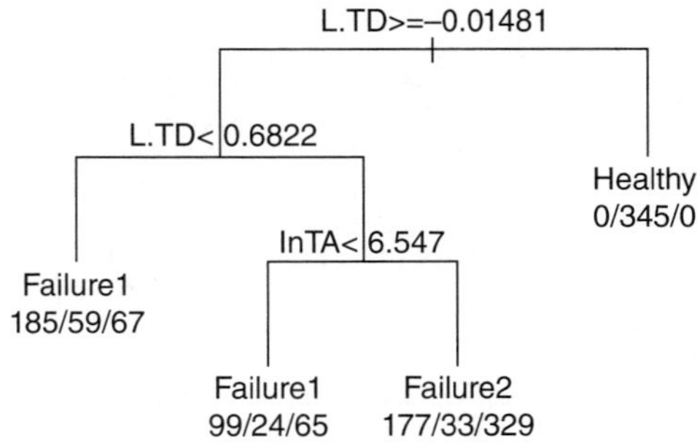

Figure 5.3 Structure of the pruned tree.

Table 5.10 Confusion matrix and errors with the pruned tree.

Predicted	**Real class**					
class	**Training**		**30.73%**	**Test**		**26.144%**
Pruned tree	**Healthy**	**Failure1**	**Failure2**	**Healthy**	**Failure1**	**Failure2**
Healthy	345	0	0	42	0	0
Failure1	83	284	132	6	33	13
Failure2	33	177	329	3	18	38

This tree obtained a training error of 30.73% and a test error of 26.144%. Analyzing the confusion matrices (Table 5.10), there was no firm in the failure1 or failure2 categories classified as healthy, which was the worst error in this problem.

Using the `adabag` library, an AdaBoost.M1 classifier was built with 1000 trees pruned using `maxdepth = 3` to limit the size of the individual tree in each boosting epoch. The higher complexity of the three-class problem required a larger ensemble and bigger trees than in the two-class case.

```
> sabi.boosting <- boosting(Class ~ ., data=sabi[ind,], mfinal=1000, maxdepth=3)
> sabi.boosting

$formula
CLASS ~ .


$trees
[[1]]
n= 1383
node), split, n, loss, yval, (yprob)
      * denotes terminal node
 1) root 1383 895 Healthy (0.34056399 0.35285611 0.30657990)
   2) L.TD>=-0.03934943 1023 552 Failure1 (0.46041056 0.12512219 0.41446725)
     4) L.TD< 0.7258538 366 150 Failure1 (0.59016393 0.22131148 0.18852459) *
     5) L.TD>=0.7258538 657 302 Failure2 (0.38812785 0.07153729 0.54033486)
      10) CF.TD>=0.1505895 22   0 Failure1 (1.00000000 0.00000000 0.00000000) *
      11) CF.TD< 0.1505895 635 280 Failure2 (0.36692913 0.07401575 0.55905512) *
   3) L.TD< -0.03934943 360   0 Healthy (0.00000000 1.00000000 0.00000000) *

[[2]]
n= 1383
node), split, n, loss, yval, (yprob)
```

```
      * denotes terminal node
 1) root 1383 909 Failure1 (0.342733189 0.328994939 0.328271873)
   2) CF.TD< 0.366587 1058 587 Failure1 (0.445179584 0.125708885 0.429111531)
     4) L.TD< 0.5145494 148  60 Failure1 (0.594594595 0.283783784 0.121621622) *
     5) L.TD>=0.5145494 910 474 Failure2 (0.420879121 0.100000000 0.479120879)
      10) NACE1=5,7,9 387 189 Failure1 (0.511627907 0.136950904 0.351421189) *
      11) NACE1=0,1,2,3,4,6,8 523 223 Failure2 (0.353728489 0.072657744 0.573613767) *
   3) CF.TD>=0.366587 325   3 Healthy (0.009230769 0.990769231 0.000000000) *
...
[[999]]
n= 1383
node), split, n, loss, yval, (yprob)
      * denotes terminal node
1) root 1383 889 Failure1 (0.357194505 0.317425886 0.325379610)
  2) CF.TD< 0.3689572 1044 551 Failure1 (0.472222222 0.096743295 0.431034483)
    4) L.TD< 0.7657824 463 181 Failure1 (0.609071274 0.127429806 0.263498920) *
    5) L.TD>=0.7657824 581 253 Failure2 (0.363166954 0.072289157 0.564543890) *
  3) CF.TD>=0.3689572 339   1 Healthy (0.002949853 0.997050147 0.000000000) *

[[1000]]
n= 1383

node), split, n, loss, yval, (yprob)
      * denotes terminal node

 1) root 1383 887 Failure1 (0.35864064 0.30079537 0.34056399)
   2) L.TD>=0.006869229 1097 601 Failure1 (0.45214221 0.11850501 0.42935278)
     4) L.TD< 0.5189651 170  70 Failure1 (0.58823529 0.28823529 0.12352941) *
     5) L.TD>=0.5189651 927 477 Failure2 (0.42718447 0.08737864 0.48543689)
      10) lnCA>=9.794014 48   7 Failure1 (0.85416667 0.02083333 0.12500000) *
      11) lnCA< 9.794014 879 435 Failure2 (0.40386803 0.09101251 0.50511945) *
   3) L.TD< 0.006869229 286   0 Healthy (0.00000000 1.00000000 0.00000000) *
$weights
  [1] 0.364578809 0.346573590 0.173074265 0.359646230 0.276544082 0.364578809
+ 0.097925705
  [8] 0.174564531 0.103769277 0.065894194 0.103769277 0.265658676 0.219662829
+ 0.049932998
 [15] 0.186514916 0.067346754 0.185018303 0.058635436 0.213602547 0.264108683
+ 0.140468342
  ...
[498] 0.399603376 0.353095422 0.315968225 0.364578809 0.111083733 0.301671643
+ 0.05428319
[505] 0.061538156 0.20002252 0.236412253 0.177547374 0.281228753 0.105231258
+ 0.118410046
[512] 0.201527734 0.134573711 0.083345015 0.191009745 0.052832906 0.140468342
+ 0.12868836
  ...
[981] 0.065894194 0.084801392 0.224218462 0.158212271 0.020971982 0.19851820
+ 0.111083733
[988] 0.023865701 0.112548021 0.130158852 0.102307737 0.204540874 0.118410046
+ 0.00100000
[995] 0.399603376 0.381145778 0.323964787 0.340079993 0.164148396 0.230306944
$votes
         [,1]         [,2]         [,3]
  [1,] 8.263977e+01   8.88095187  97.14084282
  [2,] 1.058770e+02  33.87221740  48.91229742
  [3,] 7.819811e+01  18.70088739  91.76256445
  [4,] 8.707288e+01  25.38437594  76.20430924
  [5,] 1.013104e+02  29.84165791  57.50950139
   ...
[460,] 77.07325 24.359665  87.22865
[461,] 86.47337  1.031470 101.15672
```

```
[462,] 61.79782  0.183762 126.67998
[463,] 77.64892  8.798496 102.21415
[464,] 46.79873 16.036148 125.82668
[465,] 52.79921 20.295619 115.56673
...
[920,]  86.032071  15.666665  86.9628263
[921,]  70.334356   1.256456 117.0707496
[922,]  73.072574   1.345051 114.2439367
[923,]   0.000000 188.661562   0.0000000
[924,] 100.015399  61.086916  27.5592464
[925,]   0.191526 188.293831   0.1762049
...
[1379,] 2.69594697 185.6792 0.2864602
[1380,] 0.08871523 188.5728 0.0000000
[1381,] 0.19152599 188.4700 0.0000000
[1382,] 0.00000000 188.6616 0.0000000
[1383,] 0.00100000 188.6606 0.0000000

$prob
          [,1]        [,2]      [,3]
[1,] 0.4380318 0.04707346 0.5148947
[2,] 0.5612009 0.17953958 0.2592595
[3,] 0.4144888 0.09912399 0.4863872
[4,] 0.4615295 0.13454980 0.4039207
[5,] 0.5369955 0.15817561 0.3048289
...
[460,] 0.4085265 0.1291183242 0.4623552
[461,] 0.4583518 0.0054673044 0.5361809
[462,] 0.3275592 0.0009740298 0.6714668
[463,] 0.4115778 0.0466364007 0.5417858
[464,] 0.2480565 0.0849995492 0.6669439
[465,] 0.2798621 0.1075768614 0.6125611
...
[920,] 0.456012714 0.083041106 0.4609461803
[921,] 0.372807029 0.006659840 0.6205331309
[922,] 0.387320944 0.007129438 0.6055496181
[923,] 0.000000000 1.000000000 0.0000000000
[924,] 0.530131301 0.323791002 0.1460776965
[925,] 0.001015183 0.998050843 0.0009339736
...
[1379,] 1.428986e-02 0.9841918 0.001518381
[1380,] 4.702348e-04 0.9995298 0.000000000
[1381,] 1.015183e-03 0.9989848 0.000000000
[1382,] 0.000000e+00 1.0000000 0.000000000
[1383,] 5.300497e-06 0.9999947 0.000000000
$class

> sabi.boosting$class

   [1] "Failure2" "Failure1" "Failure2" "Failure1" "Failure1" "Failure2" "Failure2"
   [8] "Failure1" "Failure1" "Failure2" "Failure1" "Failure1" "Failure1" "Failure1"
  [15] "Failure1" "Failure2" "Failure1" "Failure1" "Failure2" "Healthy" "Failure1"
  ...
 [456] "Failure2" "Failure2" "Failure2" "Failure1" "Failure2" "Failure2" "Failure2"
 [463] "Failure2" "Failure2" "Failure2" "Failure2" "Failure2" "Failure2" "Failure2"
 [470] "Failure2" "Failure2" "Failure2" "Failure2" "Failure2" "Failure2" "Failure2"
 ...
 [918] "Failure2" "Failure2" "Failure2" "Failure2" "Failure2" "Healthy""Failure1"
 [925] "Healthy" "Healthy" "Healthy" "Healthy" "Failure2" "Healthy" "Healthy"
 [932] "Failure1" "Healthy" "Healthy" "Healthy" "Healthy" "Healthy" "Healthy"
 ...
[1365] "Healthy" "Healthy" "Healthy" "Failure1" "Healthy" "Healthy" "Healthy"
[1372] "Healthy" "Healthy" "Healthy" "Healthy" "Healthy" "Healthy" "Healthy"
```

```
[1379] "Healthy"    "Healthy"    "Healthy"    "Healthy"    "Healthy"
$importance
        LE      NACE1       NI.TA    EBIT.TA   EBIT.CAP       CA.TA        C.TA
 1.6042781  6.7914439  0.8288770 14.9197861  4.4919786  2.5133690  1.9786096
     S.CAP      CA.CL        C.CL       WC.S       L.TD       CF.TD       WC.TA
 1.7647059  1.6577540  6.1497326  0.4812834 23.6096257 11.0695187  0.3475936
      S.TA       S.CA        lnTA       lnCA
 6.4438503  2.7540107  5.4010695  7.1925134
```

Figure 5.4 shows all variables arranged from greater to lesser relative importance. In this case, the most outstanding ratios were L.TD, EBIT.TA, and CF.TD, with values at this measure of 23.61%, 14.92%, and 11.07%, respectively. Those variables which were different from financial ratios (NACE1, lnCA, lnTA, and LE) had an interesting joint contribution of 20.98%.

The test error was reduced to 21.57%, so there was a reduction of 17.5% compared with the individual tree test error (26.14%). Moreover, this important reduction was achieved by keeping the number of failed firms classified as healthy at zero for both types of failure. Table 5.11 shows the error in the training and test sets of the boosting model.

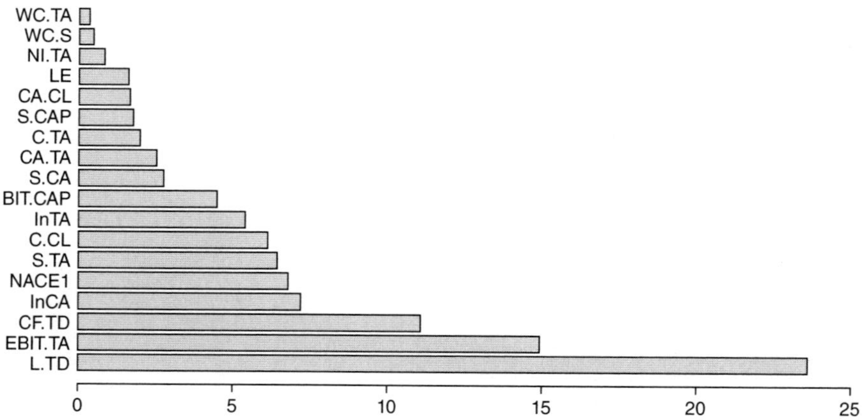

Figure 5.4 Variable relative importance in AdaBoost.M1 (three classes).

Table 5.11 Confusion matrix and errors with the AdaBoost.M1 model.

Predicted	Real class					
class	Training		24.657%	Test		21.569%
AdaBoost.M1	Healthy	Failure1	Failure2	Healthy	Failure1	Failure2
Healthy	345	3	0	42	0	0
Failure1	67	310	94	5	37	10
Failure2	29	148	367	4	14	41

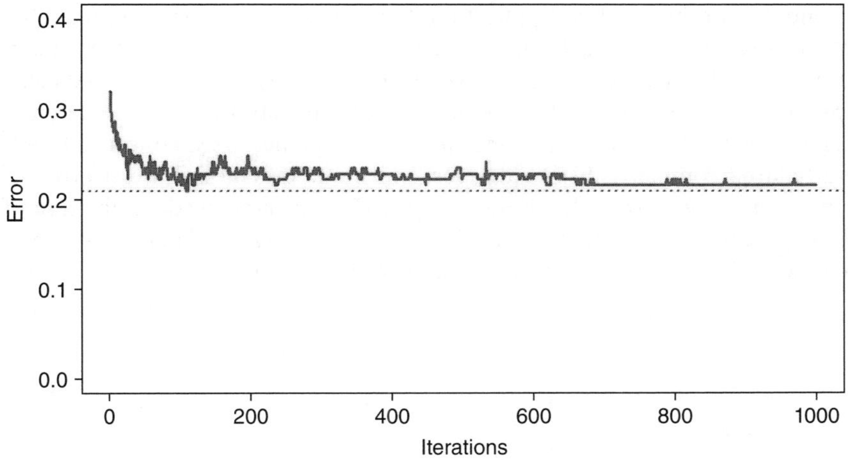

Figure 5.5 Evolution of the test error in AdaBoost.M1 (three classes).

Figure 5.5 shows the evolution of the test error as the size of the ensemble grows. This allows us to prune the ensemble to 109 trees when the minimum test error is achieved. This way we get a simpler model with a better predictive capacity (20.915%).

```
> sabi.predboosting109<- predict.boosting(sabi.boosting, newdata=sabi[-ind,],
newmfinal = 109)
> sabi.predboosting109$confusion

               Observed Class
Predicted Class Healthy Failure1 Failure2
       Healthy       42        0        0
       Failure1       5       36        8
       Failure2       4       15       43

> sabi.predboosting109$error

[1] 0.2091503
```

5.4 Conclusions

The usefulness of ensemble trees for bankruptcy prediction has been shown in this chapter. After a brief literature revision of corporate failure prediction, the application section was divided in two parts.

First, the application worked as usual with two classes, where healthy companies were distinguished from failed ones. It is more common to distinguish two classes (as in 5.3.1) than three classes (5.3.2). The legal concept of corporate failure was used, which includes bankruptcy and temporary receivership firms. Three classification methods were compared, showing the improvement in accuracy that AdaBoost achieves against LDA and Neural Network. AdaBoost achieved a test error of 8.898%, so failed firms are properly differentiated from healthy companies. This result means that the AdaBoost strategy for combining single trees achieved a reduction of 30% in the test error compared with the individual neural network and a reduction of 56.25% in comparison to the LDA test error.

In the second part of the practical application it was shown that more than one type of corporate failure can be used, including bankruptcy and temporary receivership firms (failure2), as well as acquired and dissolved firms (failure1). Therefore, failed firms were grouped in these two classes. Consequently, the application worked, differently to the usual way, with three classes, where healthy companies were contrasted with failure1 and failure2 companies. The AdaBoost.M1 method achieved a test error of 21.57% in this case (20.91% with the prune option). This result showed a reduction of 17.5% with respect to the test error of the classification tree (26.14%) and confirmed that the Adaboost.M1 algorithm outperforms individual decision trees.

6

Experiments with Adabag in Biology Classification Tasks

M. Fernández-Delgado, E. Cernadas, and M. Pérez-Ortiz

6.1 Classification of Color Texture Feature Patterns Extracted From Cells in Histological Images of Fish Ovary

The assessment of fecundity is fundamental in the study of biology and to define the management of sustainable fisheries. Stereometry (Emerson et al., 1990) is an accurate method to estimate fecundity from histological images, although it is rarely developed because it requires a lot of time from specialized technicians. Figure 6.1 shows some histological images of fish species *Merluccius merluccius* (also named European hake). The fecundity estimation requires the measurement of the diameter of matured cells (oocytes), so they must be detected and classified according to the presence or absence of nucleus, and according to the stage of development: *cortical alveoli, hydrated* or *vitellogenic/atretic*. The software Govocitos[1] (González-Rufino et al., 2013) supports technicians in fecundity studies using histological images of fish ovary, allowing oocyte detection (which can be interactively corrected by the user) and classification (according to the presence/absence of nucleus and the development stage), and using the oocyte measurements to automatically estimate the fish fecundity.

We compared 18 state-of-the-art classifiers belonging to several families for both problems (detection of cell nucleus and cell stage classification):

- **statistical classifiers:** includes linear and quadratic discriminant analysis (LDA and QDA), flexible and mixture discriminant analysis (FDA and MDA), generalized and multinomial log-linear models (GLM and MLM), and K-nearest neighbors (KNN)
- **neural networks**: multilayer perceptron (MLP), the support vector machine (SVM), using LibSVM (Chang and Lin, 2011), radial basis function (RBF) neural networks, learning vector quantization (LVQ), probabilistic neural networks (PNNs) and extreme learning machines (ELMs) (Huang et al., 2011)
- **ensembles and decision trees**: bagging (Breiman, 1996a), boosting, Adaboost.M1 (using the R package Adabag (Alfaro et al., 2007)) and random forest (Breiman, 2001) of RPART classifiers.

1 `http://citius.usc.es/w/govocitos.`

Ensemble Classification Methods with Applications in R, First Edition.
Edited by Esteban Alfaro, Matías Gámez, and Noelia García.

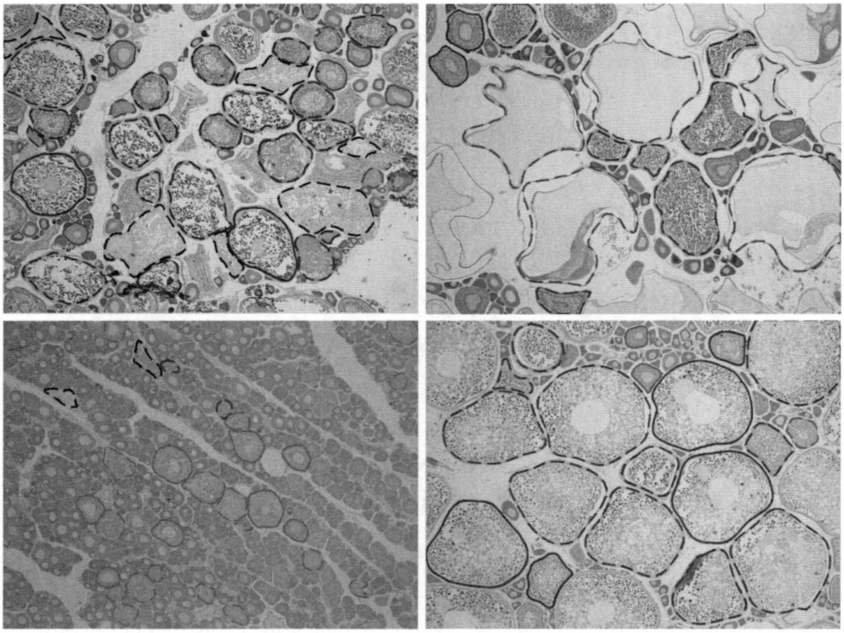

Figure 6.1 Histological images of fish species *Merluccius merluccius*, with cell outlines manually annotated by experts. The continuous (respectively dashed) lines are cells with (resp. without) nucleus. The images contain cells in the different states of development (hydrated, cortical alveoli, vitellogenic, and atretic).

These classifiers use input patterns composed by color and texture features extracted from the image. We developed an exhaustive comparison of 126 types of feature vectors, including color and texture features:

- **statistical color features:** color mean, first-order statistics, and histogram Fourier descriptors
- **grey-level texture features:** co-occurrence matrix, grey level run length, neighboring grey level dependence, sum and difference histograms, local binary patterns, and fractal analysis
- **multiscalar grey level texture features:** wavelet, multi-resolution, and pyramidal
- **integrative color texture features**.

A preliminary analysis using the SVM selected the 25 most useful feature vectors for classification, and these vectors were fed to the collection of 18 classifiers for the detection of the cell nucleus and for cell stage classification.

Figure 6.2 shows the results in terms of average accuracy and Friedman ranking (decreasing with the classifier quality) over all the feature vectors for the detection of nucleus (upper panels) and the classification of development stages (lower panels). Adabag is the sixth best nucleus detector with 79.8% accuracy against 85% for SVM (upper left panel). Considering the best accuracy achieved by each classifier on any feature vector (Figure 6.3, left panel) Adabag achieves 84.7%, just 2.5% below SVM (87.2%), and is fourth of 18 classifiers, just after SVM, MLM and RBF. For the classification of

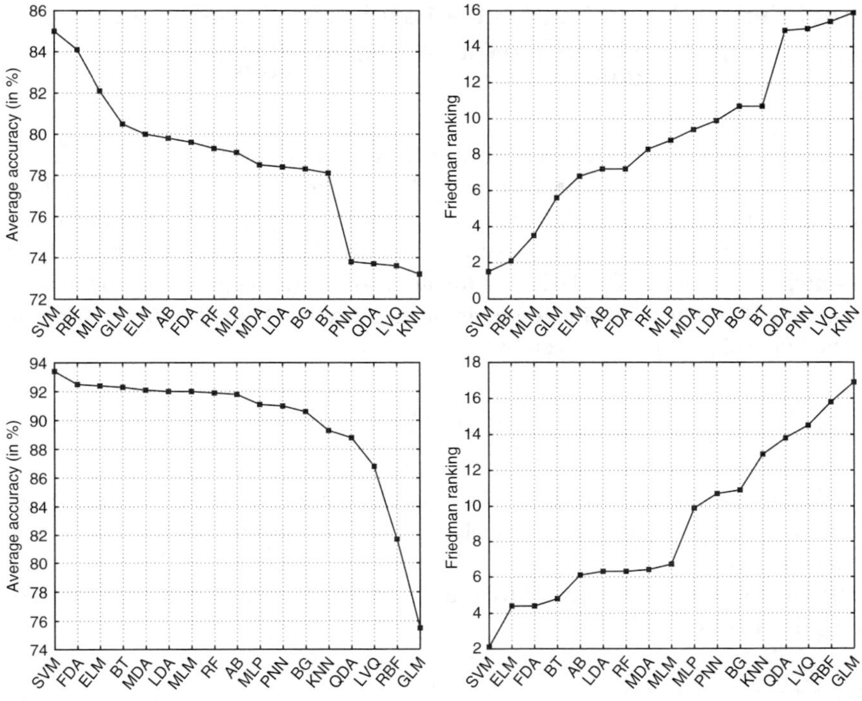

Figure 6.2 Average accuracy (in %, left panel) and Friedman rank (right panel) over all the feature vectors for the detection of the nucleus (upper panels) and stage classification (lower panels) of fish ovary cells.

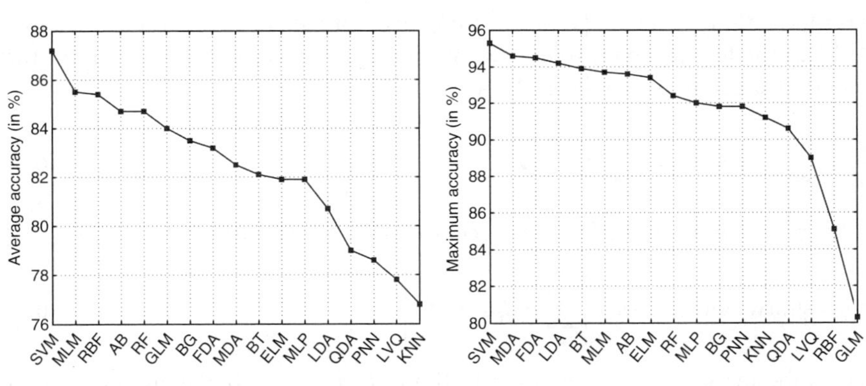

Figure 6.3 Maximum accuracies (in %), in decreasing order, achieved by the different classifiers for the detection of the nucleus (left panel) and stage classification (right panel) of fish ovary cells.

stages, Adabag is the fifth best classifier according to the Friedman ranking (Figure 6.2, lower right panel), and the ninth best classifier by average accuracy (91.8% against 93.4% with SVM, lower left panel). In addition, Adabag achieves the seventh best accuracy (Figure 6.3, right panel) in the stage classification, 93.6%, only 1.7% less than the best (SVM, 95.3%).

6.2 Direct Kernel Perceptron: Ultra-Fast Kernel ELM-Based Classification with Non-Iterative Closed-Form Weight Calculation

The direct kernel perceptron (DKP) (Fernández-Delgado et al., 2014a) is a very simple and fast kernel-based classifier, related to the SVM and the ELM (Huang et al., 2011), whose trainable parameters are calculated directly, without any iterative training, using an analytical closed-form expression that involves only the training patterns and the classes to which they belong. The DKP is a linear classifier (perceptron) in the feature space whose weight vector $\mathbf{w}$ minimizes an error measure $E(\mathbf{w})$ combining the training error and the hyperplane margin, given by:

$$E(\mathbf{w}) = \sum_{i=1}^{N} f\left(\frac{\mathbf{w}^T\mathbf{z}_i}{|\mathbf{w}|} - \gamma\right) = \mathbf{f}\left(\frac{\mathbf{Z}^T\mathbf{w}}{|\mathbf{w}|} - \gamma\vec{1}\right)^T \vec{1} \tag{6.1}$$

where $\mathbf{z}_i = d_i\mathbf{x}_i$, while $\mathbf{x}_i \in \mathbb{R}^n$ and d_i are the N training patterns and their class labels ($d_i = \pm 1$ for two-class problems), respectively, $f(x)$ is a decreasing function, the $n \times N$ matrix $\mathbf{Z}$ has the vectors $\mathbf{z}_i$ by columns, and $\gamma > 0$ is a threshold for the hyperplane margin. Deriving this error with respect to $\mathbf{w}$ and equaling to zero to minimize $E(\mathbf{w})$ we achieve the following expression:

$$\mathbf{Z}\,\mathbf{g}\left(\frac{\mathbf{Z}^T\mathbf{w}}{|\mathbf{w}|} - \gamma\vec{1}\right) = \lambda\frac{\mathbf{w}}{|\mathbf{w}|} \tag{6.2}$$

where $\mathbf{g}(\mathbf{v}) = [f'(v_1), \ldots, f'(v_n)]$, $\vec{1}$ is a vector with n ones, and λ is a free parameter because the solutions have one degree of freedom. Using a nonlinear mapping $\mathbf{\Phi}(\mathbf{x})$ associated to, for example, a Gaussian kernel with spread σ defined by $K(\mathbf{x}_1, \mathbf{x}_2) = \mathbf{\Phi}(\mathbf{x}_1)^T\mathbf{\Phi}(\mathbf{x}_2) = \exp(-|\mathbf{x}_1 - \mathbf{x}_2|^2/\sigma^2)$, we introduce a variable change from $\mathbf{w} \in \mathbb{R}^n$ in the feature space to $\vec{\alpha} \in \mathbb{R}^N$ in the input space, and the weight vector $\mathbf{w}$ can be written, analogously to the SVM, as:

$$\frac{\mathbf{w}}{|\mathbf{w}|} = \sum_{i=1}^{N} \alpha_i d_i \mathbf{\Phi}(\mathbf{x}_i) = \sum_{i=1}^{N} \alpha_i \mathbf{z}_i = \mathbf{Z}\vec{\alpha} \tag{6.3}$$

where we use $\mathbf{z}_i = d_i\mathbf{\Phi}(\mathbf{x}_i)$ in the feature space, so the classifier output is given by:

$$y(\mathbf{x}) = \text{sgn}[\mathbf{w}^T\mathbf{\Phi}(\mathbf{x})] = \text{sgn}\left[\sum_{i=1}^{N} \alpha_i d_i K(\mathbf{x}_i, \mathbf{x})\right] \tag{6.4}$$

where $\mathbf{\Phi}(\mathbf{x})$ is the mapping from the input to the feature space. Replacing the expression 6.3 in the general solution for $\mathbf{w}$ (eq. 6.2), and considering that in general there are n equations and $N > n$ unknowns $(\alpha_1, \ldots, \alpha_N)$, for a solution to exist it must be $\mathbf{g}(\mathbf{P}\vec{\alpha} - \gamma\vec{1}) = \lambda\vec{\alpha}$, where $\vec{\alpha}$ is the unknown vector and the $N \times N$ matrix $\mathbf{P}$ has elements $P_{ij} = d_i d_j \mathbf{\Phi}(\mathbf{x}_i)^T\mathbf{\Phi}(\mathbf{x}_j) = d_i d_j K(\mathbf{x}_i, \mathbf{x}_j)$. Using a linear function $f(x) = -x$, then $\mathbf{g}(\mathbf{v}) = -\vec{1}$ so that $\vec{\alpha} = \vec{1}$, leading to the linear DKP:

$$y(\mathbf{x}) = \text{sgn}\left[\sum_{i=1}^{N} d_i K(\mathbf{x}_i, \mathbf{x})\right] \tag{6.5}$$

This expression has no trainable parameter, so there is no train stage, and only the Gaussian kernel spread σ must be tuned. Its computational complexity is $\mathcal{O}(N, n)$ (linear with respect to N and n, the number of training patterns and inputs, respectively). Using a linear kernel $\mathbf{\Phi}(\mathbf{x}) = \mathbf{x}$, e.g. in high-dimensional classification problems, then $K(\mathbf{x}_i, \mathbf{x}) = \mathbf{x}_i^T\mathbf{x}$ and the weight vector can be calculated explicitly as $\mathbf{w} = \sum_i d_i\mathbf{x}_i$ and $y(\mathbf{x}) = \text{sgn}(\mathbf{w}^T\mathbf{x})$. On the contrary, with a Gaussian kernel the DKP requires the whole training set to be stored. Using a quadratic function $f(x) = (x-1)^2/2$, we achieve $\mathbf{g}(\mathbf{v}) = \mathbf{v} - \vec{1}$ and $y(\mathbf{x}) = \text{sgn}\left(\sum_i \alpha_i d_i K(\mathbf{x}, \mathbf{x}_i)\right)$, where $\vec{\alpha} = (1+\gamma)\mathbf{P}^{-1}\vec{1}$ and $P_{ij} = K(\mathbf{x}_i, \mathbf{x}_j)$. This requires the inversion of the N-square matrix $\mathbf{P}$, whose computational cost and storing requirements scale as $\mathcal{O}(N^2, n)$, so it is not recommended for medium and large data sets. For $C > 2$ classes, the one-against-one approach is recommended, using $C(C-1)/2$ two-class DKPs and assigning a test pattern to the most voted class.

We tested the linear and quadratic DKP over a wide collection of 42 data sets, 36 from the UCI Machine Learning repository (Dua and Karra Taniskidou, 2017) and six publicly available data sets[2] used by us in previous publications in the fields of biology, food technology, and fisheries management. The validation used 10 groups of training and validation sets, used for training and parameter tuning, and the results were averaged over 10 test sets. The spread σ of the Gaussian kernel was tuned with values in the range $\{2^i\}_{i=-15}^{2}$ (18 values).

A first experiment on the two-class data sets reported that the quadratic DKP, being much more expensive computationally due to the matrix inversion, reported performance similar to the linear DKP. Therefore, only the linear DKP was compared to a collection of 12 well-known classifiers in the literature over the whole data set collection:

- **LibSVM** (Chang and Lin, 2011) with a Gaussian kernel with the same tuning as DKP
- **MLP**, *nnet* R package, tuning the number of hidden neurons
- **AB**, Adaboost.M1 ensemble, *adabag* R package (Alfaro et al., 2007), with recursive partitioning (RPART) base classifiers
- **RF**, random forest ensemble, *randomForest* R package, with the same base classifiers
- **BG**, bagging ensemble, *ipred* R package
- **LDA**, *MASS* R package
- **KNN** classifier, *ipred* R package, tuning the number of neighbors between 1 and 20
- **DPP** (direct parallel perceptron), with three perceptrons
- **ELM**, Matlab code[3], tuning the number of hidden neurons and available activation functions
- **PNN** (Specht, 1990), Matlab code, tuning the Gaussian spread
- **RBF**, Matlab code, tuning the kernel spread
- **GART** (generalized ART), a hybrid between adaptive resonance theory (ART) and generalized regression neural network (GRNN).

Figure 6.4 shows the average accuracy (left panel, sorted decreasingly) and the Friedman rank (right panel, sorted increasingly) of the classifiers over all the data sets. Adaboost.M1 achieves the best average accuracy (85.3%, tied with RF) and the third Friedman rank, just after SVM (average accuracy 84.6%) and RF. Figure 6.5 shows the

2 http://persoal.citius.usc.es/manuel.fernandez.delgado/papers/dkp
3 `http://www.extreme-learning-machines.org`.

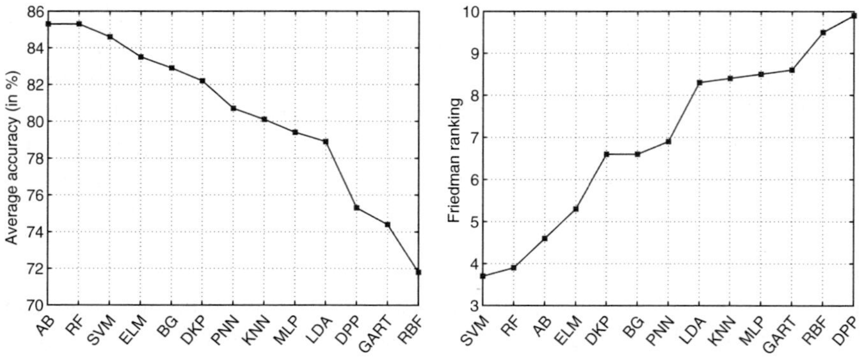

Figure 6.4 Average accuracy (left panel, in %) and Friedman ranking (right panel, decreasing with performance) of each classifier.

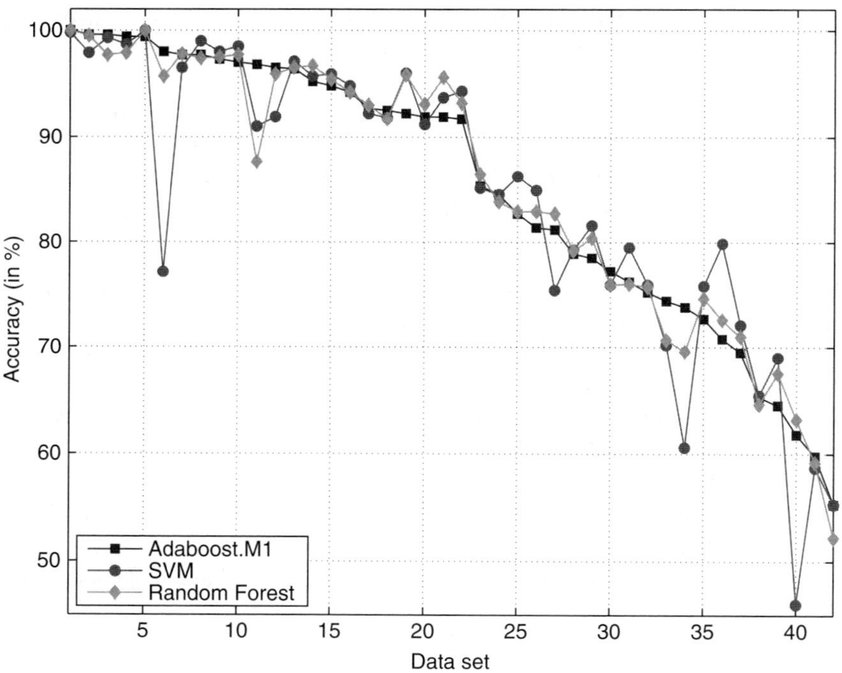

Figure 6.5 Accuracies achieved by Adaboost.M1, SVM, and random forest for each data set.

accuracies of Adaboost.M1, SVM and RF: for a few data sets SVM or RF is slightly better, but they are much worse for several data sets. The DKP achieved intermediate results: sixth position in the Friedman rank and average accuracy 82.2%. On the other hand, Figure 6.6 reports that DKP is much faster than other classifiers, such as SVM and ELM (note the logarithmic vertical scale), although this difference is lower for large data sets due to the storing of the whole training set.

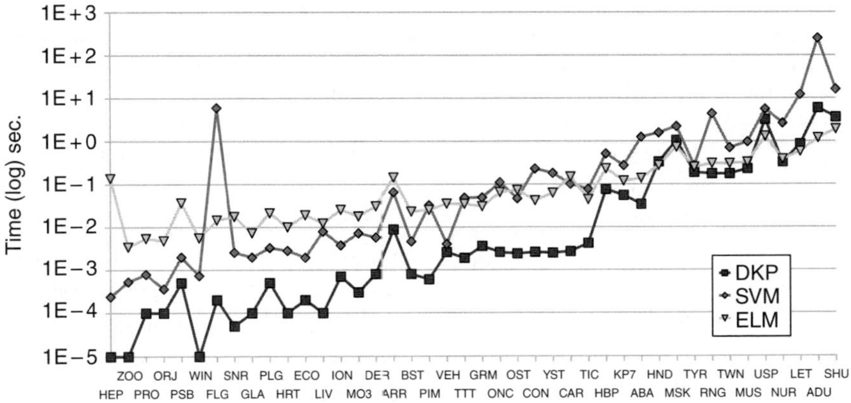

Figure 6.6 Times achieved by the faster classifiers (DKP, SVM, and ELM) for each data set, ordered by increasing size of data set.

6.3 Do We Need Hundreds of Classifiers to Solve Real-World Classification Problems?

The objective of this work (Fernández-Delgado et al., 2014b) is to develop an exhaustive comparison of a representative sample of the currently available classifiers using a very large and diverse collection of data sets in order to know which classifier is expected to give the best result (or, at least, a result not much different to the best attainable result) for a new classification problem. When a researcher or data analyzer faces the classification of a data set, he/she usually applies the classifier which he/she expects to be "the best one". This expectation may be conditioned by the (often partial) researcher knowledge about the available classifiers. Some classifiers may work very well, but they are not applied because they are provided from some field which is unknown to the researcher. The lack of available implementations for many classifiers is a major drawback, although it has been partially reduced due to the large number of classifiers implemented in Matlab, in Weka,[4] and in R[5], especially using the caret package (Kuhn, 2008), which provides a very easy interface for the execution of many classifiers, allowing automatic parameter tuning and providing a set of reasonable values for them. Another drawback is the lack of comparisons in the literature among classifiers coming from different fields: usually the papers which propose a new classifier compare it only to classifiers within the same family, excluding families outside the author's area of expertise, so we do not have an idea of the comparative performances of classifiers from different families. On the other hand, since the results of each classifier change among data sets, it is also important to develop comparisons using a very large and diverse collection of data sets to avoid arriving at conclusions that may be biased by the data sets selected.

We evaluate **179 classifiers** arising from **17 families**: discriminant analysis, Bayesian classifiers, neural networks, SVMs, decision trees, rule-based classifiers, boosting (including Adaboost.M1 provided by the Adabag package), bagging, stacking, random forests, other ensembles, GLMs, nearest-neighbors, partial least squares (PLS) and

4 See http://www.cs.waikato.ac.nz/ml/weka.
5 See http://www.r-project.org.

Table 6.1 Collection of 121 data sets from the UCI database and our real-world problems. ac-inflam, acute inflammation; bc, breast cancer; congress-voting, congressional voting; ctg, cardiotocography; conn-bench-sonar, connectionist benchmark sonar mines rocks; conn-bench-vowel, connectionist benchmark vowel deterding; pb, Pittsburg bridges; st, statlog; vc, vertebral column.

abalone	energy-y1	monks-2	spambase
ac-inflam	energy-y2	monks-3	spect
acute-nephritis	fertility	mushroom	spectf
adult	flags	musk-1	st-australian-credit
annealing	glass	musk-2	st-german-credit
arrhythmia	haberman-survival	nursery	st-heart
audiology-std	hayes-roth	oocMerl2F	st-image
balance-scale	heart-cleveland	oocMerl4D	st-landsat
balloons	heart-hungarian	oocTris2F	st-shuttle
bank	heart-switzerland	oocTris5B	st-vehicle
blood	heart-va	optical	steel-plates
breast-cancer	hepatitis	ozone	synthetic-control
bc-wisc	hill-valley	page-blocks	teaching
bc-wisc-diag	horse-colic	parkinsons	thyroid
bc-wisc-prog	ilpd-indian-liver	pendigits	tic-tac-toe
breast-tissue	image-segmentation	pima	titanic
car	ionosphere	pb-MATERIAL	trains
ctg-10classes	iris	pb-REL-L	twonorm
ctg-3classes	led-display	pb-SPAN	vc-2classes
chess-krvk	lenses	pb-T-OR-D	vc-3classes
chess-krvkp	letter	pb-TYPE	wall-following
congress-voting	libras	planning	waveform
conn-bench-sonar	low-res-spect	plant-margin	waveform-noise
conn-bench-vowel	lung-cancer	plant-shape	wine
connect-4	lymphography	plant-texture	wine-quality-red
contrac	magic	post-operative	wine-quality-white
credit-approval	mammographic	primary-tumor	yeast
cylinder-bands	miniboone	ringnorm	zoo
dermatology	molec-biol-promoter	seeds	
echocardiogram	molec-biol-splice	semeion	
ecoli	monks-1	soybean	

principal component regression, logistic and multinomial regression, multiple adaptive regression splines (MARS) and other methods. We used the 84 classifiers provided by Weka v. 3.6.8, 37 classifiers implemented in R, 49 in R through the *caret* R package, 6 classifiers provided by C/C++ libraries and 5 implemented in Matlab, including all the relevant classifiers available today.

We tried to select the best classifier globally for the selected data set collection, ranking each classifier and family according to its accuracy. To avoid bias in the result of

Table 6.2 Friedman ranking, average accuracy and Cohen κ (both in %) for the 30 best classifiers, ordered by increasing Friedman ranking. BG, bagging; MAB, MultiBoostAB; RC, RandomCommittee.

Rank	Accuracy	κ	Classifier	Rank	Accuracy	κ	Classifier
32.9	82.0	63.5	parRF_t (RF)	50.1	80.9	61.6	RRF_t (RF)
33.1	**82.3**	**63.6**	rf_t (RF)	51.6	80.7	61.4	RRFglobal_t (RF)
36.8	81.8	62.2	svm_C (SVM)	52.5	80.6	58.0	MAB_LibSVM_w (BST)
38.0	81.2	60.1	svmPoly_t (SVM)	52.6	79.9	56.9	LibSVM_w (SVM)
39.4	81.9	62.5	rforest_R (RF)	57.6	79.1	59.3	adaboost_R (BST)
39.6	82.0	62.0	elm_kernel_m (NNET)	58.5	79.7	57.2	pnn_m (NNET)
40.3	81.4	61.1	svmRadialCost_t (SVM)	58.9	78.5	54.7	cforest_t (RF)
42.5	81.0	60.0	svmRadial_t (SVM)	59.9	79.7	42.6	dkp_C (NNET)
42.9	80.6	61.0	C5.0_t (BST)	60.4	80.1	55.8	gaussprRadial_R (OM)
44.1	79.4	60.5	avNNet_t (NNET)	60.5	80.0	57.4	RandomForest_w (RF)
45.5	79.5	61.0	nnet_t (NNET)	62.1	78.7	56.0	svmLinear_t (SVM)
47.0	78.7	59.4	pcaNNet_t (NNET)	62.5	78.4	57.5	fda_t (DA)
47.1	80.8	53.0	BG_LibSVM_w (BAG)	62.6	78.6	56.0	knn_t (NN)
47.3	80.3	62.0	mlp_t (NNET)	62.8	78.5	58.1	mlp_C (NNET)
47.6	80.6	60.0	RotationForest_w (RF)	63.0	79.9	59.4	RC_w (OEN)

the comparison we used all the classification data sets from the UCI Machine Learning repository (121 data sets, including a few own real-world problems, publically available,[6] Table 6.1) up to a certain size to avoid execution problems. Thus, the data set collection is very wide and diverse, which should allow conclusions to be drawn about the classifier behavior that are not dependent on the selected data. The large number of classifiers (179) suggests that the highest performance achieved by some classifier and data set is very similar to the highest attainable performance for that data set, so any suboptimal result is only due to classifier limitations, not data set difficulty. The metaparameters were tuned using values which are supposed to be adequate to guarantee that all the classifiers are properly configured to work at their best performance, despite the obvious lack of experience for some classifiers. Finally, we use the same data partitioning for all the classifiers, which avoids possible bias in the comparison results.

The classifiers most likely to be the best (Table 6.2) are the RF versions (Breiman, 2001), the best of which (accessed via the *caret* R package) achieves 94.1% maximum accuracy and over 90% in 84.3% of the data sets. The RF is not significantly better (according to a T-test) than the Gaussian LibSVM (92.3% maximum accuracy). A few models are clearly better than the remaining ones: RF, Gaussian and polynomial LibSVM, Gaussian ELM, C5.0 and avNNet (a committee of MLPs provided by the *caret* R package). Figure 6.7 shows the sorted Friedman ranking and average accuracy achieved by the 25 best classifiers: Adaboost.M1 (using the Adabag package, labeled adaboost-R) has the 20th best rank of 179 classifiers, achieving 84.25% maximum accuracy, although it does not appear in the lower panel because its accuracy is below 25th position. Figure 6.8 (upper panel) shows the rank intervals (minimum, mean, and maximum) of the classifiers of the 17

6 `http://persoal.citius.usc.es/manuel.fernandez.delgado/papers/jmlr/data.tar.gz`.

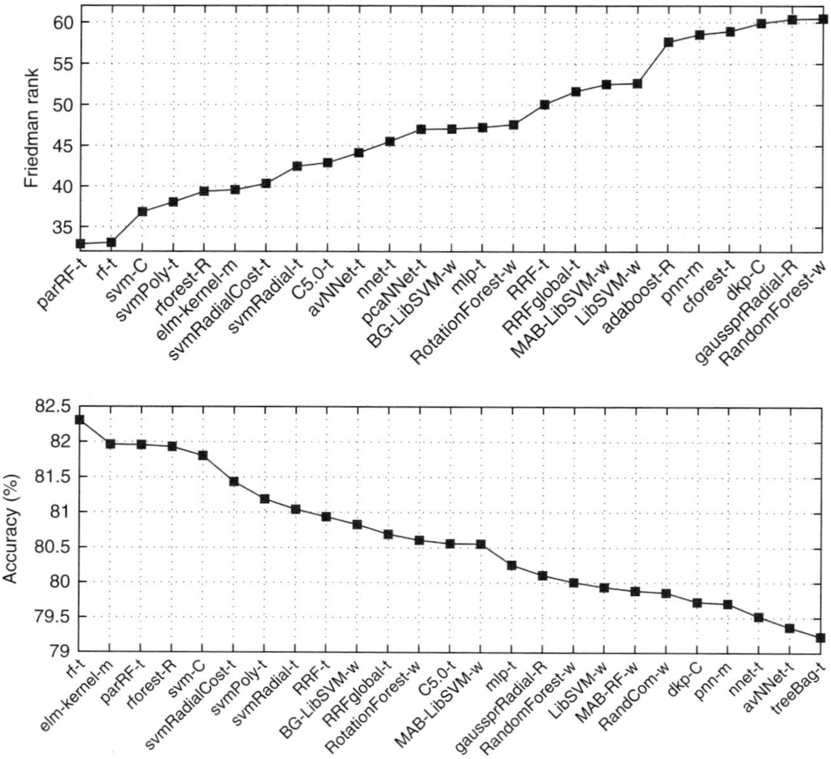

Figure 6.7 Friedman rank (upper panel, increasing order) and average accuracies (lower panel, decreasing order) for the 25 best classifiers.

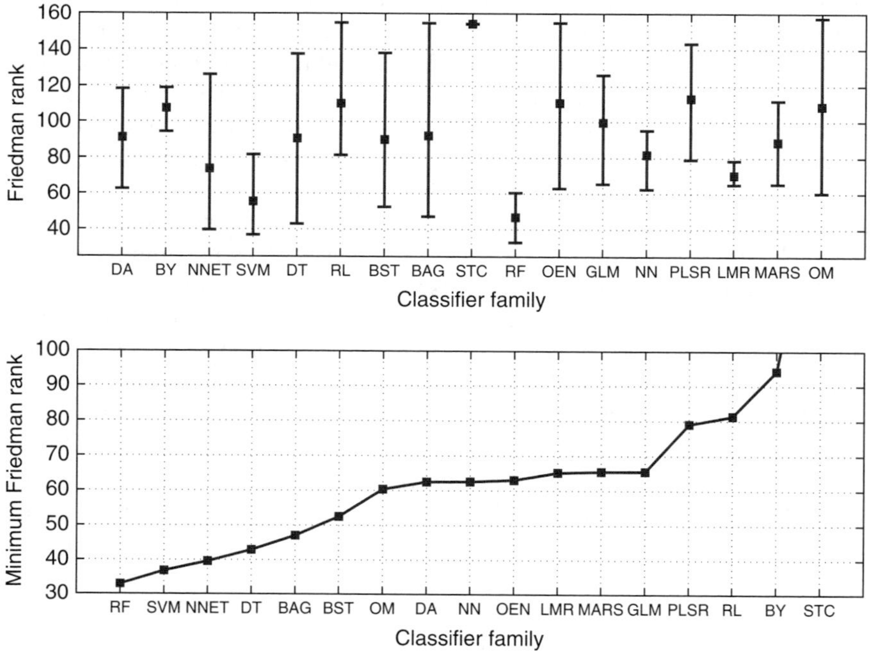

Figure 6.8 Friedman rank interval for the classifiers of each family (upper panel) and minimum rank (by ascending order) for each family (lower panel).

families. The lower panel sorts the 17 families by the best rank achieved by one classifier of that family. The RF family is the best (three out of the five best classifiers), followed by SVM (four classifiers in the top 10), neural networks (NNET, five members in the top 20), decision trees (DT), bagging (BAG), and boosting (BST).

6.4 On the use of Nominal and Ordinal Classifiers for the Discrimination of Stages of Development in Fish Oocytes

The relevance of fecundity studies is very important for fisheries management. An accurate fish fecundity estimation must only consider mature oocytes, which must be reliably classified, according to their stage of development, by experienced personnel using histological images. The main developmental stages of oocytes are *primary growth* (PG; immature), *cortical alveoli*, *vitellogenic* and *hydrated* (CA, VIT and HYD, respectively; mature stages), and *atretic* (AT; resorpted or non-ovulated oocytes). For species *Reinhardtius hippoglossoides* (RH, also called Greenland halibut) the VIT stage can be divided into four substages: VIT1, VIT2, VIT3, and VIT4 (Figure 6.9). A correct classification of the VIT substages is necessary to estimate egg production and calculate fertility. In the current paper we use ordinal classifiers to fully capture the temporal evolution of stages (Pérez-Ortiz et al., 2015) by minimizing error measures specific to ordinal classification and compare them to the state-of-the-art nominal classifiers.

The literature defines several groups of ordinal classifiers. The *threshold methods* include **POM** (proportional odds model) (McCullagh and Nelder, 1989), a reformulation

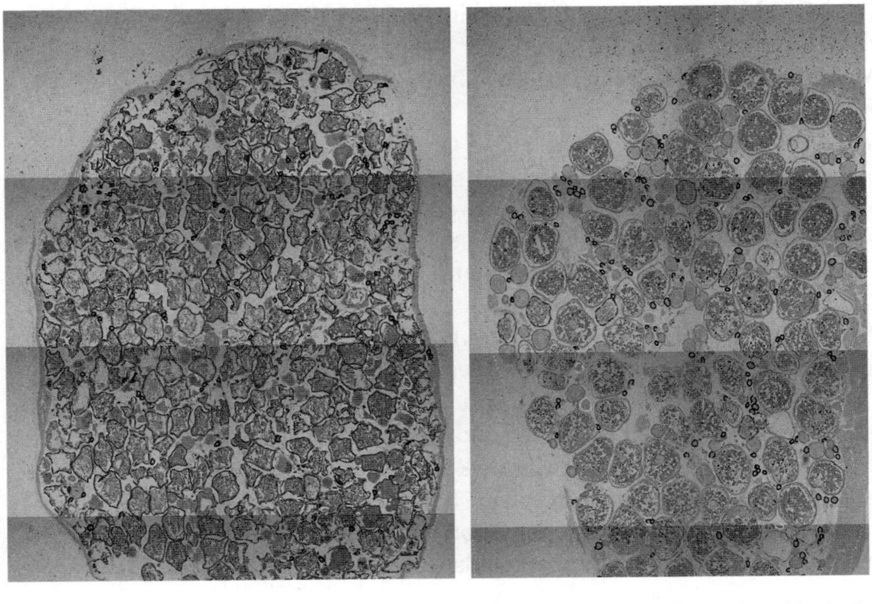

Figure 6.9 Examples of histological images of fish species *Reinhardtius hippoglossoides*, including oocytes with the six different development stages (PG, CA, VIT1, VIT2, VIT3, and VIT4).

of logistic regression for ordinal classification, **KDLOR** (kernel discriminant learning ordinal regression) (Sun et al., 2010), the ordinal version of discriminant learning, and **SVORIM** (support vector ordinal regression with implicit constraints), an ordinal formulation (Chu and Keerthi, 2007) of the SVM. Another group is the *decomposition methods*: **SVMOD** (SVM with ordinal decompositions), which combines decomposed labels (Frank and Hall, 2001), weighted patterns and SVM base methodology, **ELMOR** (extreme learning machine for ordinal regression) (Deng et al., 2010), and **EPSVM** (ensemble learning for ordinal regression with product combiner and SVM) (Pérez-Ortiz et al., 2014). The *reduction methods* include **REDSVM** (REDuction SVM) (Li and Lin, 2007), which extends the training data to develop an order test using specific misclassification weights and binary classifiers, and **ONN** (ordinal neural network) (Cardoso and da Costa, 2007), a similar data replication method based on an MLP. Finally, the *ensemble-based techniques* include **ORBoost** (ordinal regression boosting) and **ORBoostP** (ordinal regression boosting using perceptrons) (Lin and Li, 2006), which use MLP or single perceptron base learners. We also tried **SVR** (support vector regression). All the SVM-based ordinal or nominal methods use LibSVM (Chang and Lin, 2011) and their tunable metaparameters follow the LibSVM recommendations.

The nominal classifiers include 13 of the best methods in Fernández-Delgado et al. (2014b):

- **ABR**, Adaboost-M1 in the *Adabag* R package Alfaro et al. (2007)
- **ABW**, Adaboost.M1 in Weka
- **AvNN**: ensemble of five MLP neural networks in the *caret* R package
- **BAG**, bagging ensemble in R
- **ELM** and **GELM**, extreme learning machine in Matlab without and with Gaussian kernel
- **GSVM**, LibSVM with Gaussian kernel in C
- **LBR**, LogitBoost in R
- **LBW**, LogitBoost in Weka
- **LDA**, linear discriminant analysis in R
- **MLP-T**, multilayer perceptron neural network, *caret* R package
- **MLP-R**, *RSNNS* R package
- **MLP-PSO**, multilayer perceptron trained using particle swarm optimization (PSO) with Matlab PSO Research Toolbox[7]
- **MLRM**, multinomial logistic regression model in Weka
- **RF**, random forest in R.

The fish oocytes were manually drawn and labelled with the development stage by expert technicians of the Institute of Marine Research CSIC[8] using Govocitos software. The classifier input vectors contain 25 values: 10 grey level texture features (local binary patterns (Ojala et al., 2002), taking the uniform patterns with radius $R = 1$ and eight neighbors), and 15 chromatic features (statistical features about each RGB channel:

7 http://www.georgeevers.org/pso_research_toolbox.htm.
8 http://iim.csic.es.

mean, variance, third and fourth statistical moments and entropy). The data[9] include patterns of species:

- *Merluccius merluccius* (MC): 1022 patterns with three stages of development: CA (25.3% of the total patterns), HYD (6.0%) and VIT/AT (68.7%)
- *Trisopterus luscus* (TL): 912 patterns with stages: CA (57.6%), HYD (1.5%) and VIT/AT (40.9%)
- *Reinhardtius hippoglossoides* (RH): 16 images (one image per individual, Figure 6.1) with 7915 cells and six stages: PG (37.6%), CA (18.8%), VIT1 (20.8%), VIT2 (11.6%), VIT3 (8.0%) and VIT4 (3.2%).

With species MC and TL we used a 5-fold cross-validation, tuning the classifier meta-parameters using equal-sized training and validation sets. With species RH we developed two experiments:

- **leave-one-image-out** (LOIO) validation, with 16 trials, excluding one image for test in each trial, and tuning the meta-parameters to minimize the MAE
- **mixed images**, randomly selecting 10 trios of training, validation, and test sets, with similar parameter tuning.

The performance measures are accuracy, Cohen κ, which discards the probability of success by chance, mean absolute error (MAE), Kendall rank correlation coefficient (τ), and Spearman rank correlation coefficient (ρ).

Table 6.3 (resp. 6.4) reports the results for species MC and TL (resp. RH with LOIO and MIX methodologies), showing the best value in bold and the second best one in italic. The results are good for species MC (accuracy 94.1%, MAE 0.110 with SVMOD, ordinal) and TL (accuracy 95.6%, MAE 0.087 with AvNNet, nominal). These MAE values mean that each stage is misclassified, with neighbor stages less than 10%. For species RH, the accuracy values are worse (67.8% with SVMOD and 80.4% with GSVM for LOIO and MIX experiments), as are κ and MAE; τ and ρ are worse compared to MC, but better than TL, so the classification errors seem to keep the proper order. The LOIO are more realistic than the MIX results (whose mean accuracy and κ are 11 and 21 points higher) because the former does not include information about the test fish image in the training set. The results of ABR (Adaboost.M1 using Adabag) are not very different from the best ones for the species MC (accuracy 93.1% against 94.1% with SVMOD), TL (94.5% against 95.6% with AvNNet) and RH for the MIX experiment (77.8% against 80.4% with GSVM), but they are very different for the TL and LOIO experiments (51.8% against 67.8% with SVMOD). The ABW (Weka version of Adaboost.M1) is much worse than the Adabag version in all the species and experiments. Ordinal classifiers usually perform better in the mean in terms of MAE, τ and ρ, sometimes at the expense of lower accuracy and κ values, and the difference between them rises with the number of classes (from MC and TL to RH). With three stages, four of six possible errors (non-diagonal elements in the square confusion matrix) respect the stage ordering. However, with six stages (species RH) 20 of 30 possible errors involve non-neighbor stages.

Table 6.5 reports the average confusion matrix, sensitivities (Se) and positive predictivities (PP) achieved by SVMOD and GSVM on species RH with LOIO experiments. The diagonal values are the highest in each row and column, and only the PP of stage

9 `https://citius.usc.es/datasets/fish-ovary`.

Table 6.3 Classification results: accuracy, Cohen κ (both in %), mean absolute error (MAE), Kendall τ, and Spearman ρ for species MC and TL with three stages.

	Merluccius merluccius					Trisopterus luscus				
Classifier	**Accuracy**	κ	**MAE**	τ	ρ	**Accuracy**	κ	**MAE**	τ	ρ
POM	87.8	73.1	0.164	0.800	0.834	92.0	83.7	0.140	0.856	0.865
KDLOR	85.4	71.1	0.160	0.832	0.870	84.8	72.9	0.166	0.854	0.894
SVORIM	89.5	78.6	0.132	0.839	0.874	94.7	89.1	0.095	0.902	0.908
SVMOD	**94.1**	**86.9**	**0.110**	**0.856**	0.858	94.9	89.7	0.097	0.900	0.901
ELMOR	93.2	85.4	0.116	0.848	0.859	91.5	83.0	0.152	0.842	0.850
EPSVM	89.0	74.0	0.161	0.777	0.828	92.3	84.2	0.143	0.855	0.858
REDSVM	89.1	77.0	0.140	0.834	0.861	93.8	87.4	0.112	0.886	0.890
ONN	86.8	73.6	0.162	0.810	0.853	91.5	83.8	0.112	0.889	0.911
ORBoost	90.1	79.0	0.124	*0.854*	**0.882**	93.1	86.5	0.096	0.905	0.922
ORBoostP	90.2	78.9	0.125	0.851	*0.878*	93.3	86.7	0.098	0.902	**0.917**
SVR	84.8	69.7	0.186	0.792	0.818	84.2	71.4	0.190	0.832	0.862
Mean	89.1	77.0	**0.144**	**0.827**	**0.856**	**91.5**	**83.5**	**0.127**	**0.875**	**0.889**
ABR	93.1	85.1	0.130	0.832	0.835	94.5	88.8	0.109	0.887	0.888
ABW	81.4	59.1	0.309	0.671	0.688	79.8	56.6	0.388	0.610	0.615
AvNN	93.6	86.1	*0.115*	0.851	0.854	**95.6**	**91.1**	*0.088*	*0.909*	0.910
BAG	91.8	81.8	0.148	0.806	0.811	91.0	81.5	0.179	0.815	0.815
ELM	93.5	85.8	0.119	0.845	0.849	94.6	89.0	0.105	0.892	0.893
GELM	*93.8*	86.4	0.119	0.845	0.847	95.2	90.1	0.093	0.904	0.906
GSVM	92.8	84.2	0.134	0.823	0.826	*95.5*	*90.9*	**0.087**	**0.911**	*0.912*
LBR	91.4	82.3	0.164	0.790	0.795	92.6	87.1	0.148	0.830	0.835
LBW	91.1	80.7	0.162	0.789	0.794	90.3	80.3	0.188	0.805	0.807
LDA	92.5	83.8	0.128	0.829	0.835	93.1	85.9	0.134	0.861	0.863
MLP-T	*93.8*	*86.5*	0.129	0.834	0.837	**95.6**	**91.1**	0.088	*0.909*	0.910
MLP-R	91.6	81.6	0.156	0.800	0.804	94.3	88.4	0.113	0.883	0.883
MLP-PSO	72.5	48.2	0.308	0.688	0.722	68.2	47.3	0.395	0.633	0.668
MLRM	93.3	85.4	0.127	0.835	0.837	93.4	86.8	0.127	0.869	0.871
RF	93.1	84.9	0.128	0.831	0.835	93.5	86.7	0.133	0.862	0.862
Mean	**90.6**	**80.1**	0.158	0.805	0.811	91.1	82.8	0.158	0.839	0.843

Table 6.4 Classification results for the species RH with six stages using the LOIO and MIX methodologies.

	Leave-one-image-out					Mixed images				
Classifier	**Accuracy**	κ	**MAE**	τ	ρ	**Accuracy**	κ	**MAE**	τ	ρ
POM	63.9	49.5	0.394	0.770	0.851	70.5	64.6	0.310	0.879	0.940
KDLOR	66.7	52.9	0.360	0.789	0.856	78.9	74.7	0.222	0.912	*0.957*
SVORIM	67.2	53.5	*0.354*	0.788	0.858	79.0	74.8	0.222	0.911	*0.957*
SVMOD	**67.8**	**54.6**	**0.352**	*0.796*	**0.861**	*80.1*	*76.2*	*0.215*	0.912	*0.957*
ELMOR	62.7	49.0	0.418	0.762	0.831	77.5	73.0	0.247	0.898	0.948
EPSVM	66.9	53.5	0.358	**0.798**	*0.860*	75.6	70.7	0.259	0.897	0.950
REDSVM	67.2	53.4	0.362	0.789	0.856	78.7	74.4	0.225	0.910	*0.957*
ONN	63.7	49.4	0.393	0.768	0.851	74.3	69.1	0.276	0.890	0.945
ORBoost	64.5	50.2	0.376	0.772	0.849	77.6	73.1	0.232	0.909	*0.957*
ORBoostP	64.4	50.2	0.378	0.772	0.848	77.5	73.0	0.233	0.909	0.956
SVR	66.1	52.3	0.363	0.785	0.855	79.1	74.9	0.218	**0.914**	**0.959**
Mean	**65.6**	**51.7**	**0.373**	**0.781**	**0.852**	**77.2**	**72.6**	**0.242**	**0.904**	**0.953**
ABR	51.8	34.4	0.597	0.748	0.799	77.8	73.3	0.255	0.891	0.941
ABW	30.8	10.5	0.895	0.572	0.619	32.8	19.4	0.771	0.710	0.805
AvNN	53.0	37.8	0.626	0.772	0.822	79.3	75.2	0.229	0.905	0.952
BAG	50.6	35.0	0.675	0.714	0.771	70.7	64.8	0.382	0.824	0.889
ELM	64.4	50.2	0.424	0.750	0.808	78.3	73.9	0.249	0.894	0.942
GELM	65.6	51.7	0.400	0.766	0.821	*80.1*	76.1	0.221	0.908	0.953
GSVM	*67.4*	*53.7*	0.363	0.790	0.647	**80.4**	**76.5**	**0.212**	*0.913*	0.785
LBR	45.7	33.1	0.801	0.600	0.663	71.6	70.9	0.433	0.801	0.865
LBW	56.3	40.6	0.592	0.679	0.737	72.3	66.8	0.338	0.851	0.912
LDA	53.5	39.0	0.578	0.785	0.839	77.5	72.9	0.248	0.897	0.948
MLP-T	52.3	36.0	0.599	0.768	0.824	79.0	74.9	0.234	0.903	0.950
MLP-R	51.4	35.1	0.613	0.769	0.827	76.8	72.2	0.265	0.888	0.940
MLP-PSO	32.0	16.6	0.907	0.516	0.582	36.7	24.1	0.757	0.700	0.816
MLRM	66.1	53.0	0.372	0.794	0.848	78.3	73.9	0.237	0.903	0.951
RF	52.6	36.8	0.639	0.748	0.802	78.2	73.8	0.254	0.891	0.940
Mean	52.9	37.6	0.605	0.718	0.761	71.3	65.9	0.349	0.859	0.906

Table 6.5 Confusion matrices and sensitivities/positive predictivities for each stage (in %) achieved by SVMOD and GSVM for species RH and LOIO experiments.

SVMOD-LOIO	PG	CA	VIT1	VIT2	VIT3	VIT4	Se(%)	PP(%)
PG	**22.29**	10.15	0.94	0.37	0.03	0.00	66.0	80.6
CA	5.00	**11.32**	2.29	0.09	0.01	0.00	60.5	45.3
VIT1	0.37	3.47	**14.23**	2.12	0.05	0.10	70.0	74.0
VIT2	0.00	0.04	1.68	**7.16**	0.92	0.15	71.9	68.3
VIT3	0.00	0.01	0.08	0.60	**6.00**	2.14	68.0	71.0
VIT4	0.00	0.00	0.02	0.14	1.44	**6.78**	80.9	74.0
GSVM-MIX	PG	CA	VIT1	VIT2	VIT3	VIT4	Se(%)	PP(%)
PG	**24.51**	7.66	1.17	0.29	0.14	0.00	72.5	80.3
CA	5.41	**10.36**	2.79	0.15	0.01	0.00	55.4	48.4
VIT1	0.58	3.35	**14.29**	2.02	0.03	0.08	70.2	71.6
VIT2	0.01	0.04	1.63	**7.00**	1.05	0.22	70.3	64.3
VIT3	0.00	0.00	0.05	1.23	**4.68**	2.88	53.0	62.5
VIT4	0.00	0.00	0.03	0.19	1.57	**6.59**	78.6	67.5

CA is below 50%, due to the overlap between the stages. The only high non-diagonal values are adjacent to the diagonal, corresponding to patterns assigned to a stage neighbor to the right one. The GSVM achieves higher values than SVMOD outside the diagonal, excepting the (PG,CA) and (PG,VIT1) values, learning the ordinal information with less accuracy. In addition, GSVM achieves lower Se and PP for all stages except PG because it assigns more CA patterns than SVMOD to class PG. In fact, the sensitivity of SVMOD is above 60% for all stages, while GSVM is below 56% for stages CA and VIT3. Regarding PP, the SVMOD outperforms GSVM in all the states except CA, and the difference between them is especially high in stages VIT1–VIT3.

7

Generalization Bounds for Ranking Algorithms

W. Rejchel

7.1 Introduction

We consider the ranking problem, which is popular in the machine learning community. The goal is to predict or guess the ordering between objects on the basis of their observed features. This problem has numerous applications in practice, for instance in information retrieval, banking, quality control, and survival analysis. Recently many authors have focused their attention on this subject (Freund et al., 2003; Agarwal et al., 2005; Cossock and Zhang, 2006; Rudin, 2006; Clémençon et al., 2005, 2008; Rejchel, 2012). We start this chapter with a description of the statistical framework of the ranking problem.

Let us consider two objects that are randomly selected from a population. We assume that they are described by a pair of independent and identically distributed (with respect to the measure P) random vectors $Z_1 = (X_1, Y_1)$ and $Z_2 = (X_2, Y_2)$, taking values in $\mathcal{X} \times \mathbb{R}$, where $\mathcal{X}$ is a Borel subset of $\mathbb{R}^d$. Random vectors X_1 and X_2 are regarded as observations, while Y_1 and Y_2 are unknown variables that define the ordering between objects, namely the object z_1 is better (faster, stronger etc.) than the object z_2, if $y_1 > y_2$. Our task is to predict the ordering between objects only on the basis of observations X_1, X_2. To do this we construct functions $f : \mathcal{X} \times \mathcal{X} \to \mathbb{R}$, called ranking rules, that predict the ordering between objects in the following way:

$$\text{if } f(x_1, x_2) > 0, \text{ then we predict that } y_1 > y_2.$$

The natural approach is to look for a function f which, for a fixed loss function $\psi : \mathbb{R} \to [0, \infty)$, minimizes the risk

$$Q(f) = \mathbf{E}\psi[\text{sign}(Y_1 - Y_2)f(X_1, X_2)] \tag{7.1}$$

in some family of ranking rules $\mathcal{F}$, where $\text{sign}(t) = 1$ for $t > 0$, $\text{sign}(t) = -1$ for $t < 0$ and $\text{sign}(t) = 0$ for $t = 0$. We cannot minimize (7.1) directly, but having a learning sample of independent and identically distributed (with respect to the measure P) random vectors $Z_1 = (X_1, Y_1), \ldots, Z_n = (X_n, Y_n)$ we can minimize the empirical analog of the risk (7.1). In the ranking problem a natural candidate for the empirical risk is a U-statistic

$$Q_n(f) = \frac{1}{n(n-1)} \sum_{i \neq j} \psi_f(Z_i, Z_j), \tag{7.2}$$

Ensemble Classification Methods with Applications in R, First Edition.
Edited by Esteban Alfaro, Matías Gámez, and Noelia García.

where $\psi_f(z_1, z_2) = \psi[\text{sign}(y_1 - y_2)f(x_1, x_2)]$. The sum in (7.2) is taken over all pairs of distinct indices $(i,j) \in \{1, \ldots, n\}^2$. The ranking rule $f_n = \arg\min_{f \in \mathcal{F}} Q_n(f)$ can be used as an estimator of the unknown function $f^* = \arg\min_{f \in \mathcal{F}} Q(f)$.

The $0 - 1$ loss function $\psi(t) = \mathbb{I}_{(-\infty,0)}(t)$ seems to be the most adequate, and the risk (7.1) becomes the probability of incorrect ranking. However, in this case the empirical risk (7.2) is discontinuous and its minimization is computationally difficult. To overcome this problem one usually replaces the discontinuous loss function by its convex analog. This trick has been successfully used and has allowed many popular procedures to be invented, for instance boosting algorithms (Freund and Schapire, 1997) and support vector machines (Vapnik, 1998). Thus, from now on we assume that the loss function ψ is convex.

In the paper we investigate properties of the ranking estimator f_n by studying its excess risk $Q(f_n) - Q(f^*)$, which is the difference between the risk of the estimator and the risk of the best function in the class $\mathcal{F}$. We construct generalization bounds for the excess risk that are probabilistic inequalities bounding the excess risk by an expression depending only on the size of the sample n, a loss function ψ, and a family $\mathcal{F}$. Such results for ranking estimators can be found in Clémençon et al. (2005, 2008). In these papers one obtains bounds for the excess risk with the rate $\frac{1}{\sqrt{n}}$, if the capacity of the class $\mathcal{F}$ is restricted. These results were improved in Rejchel (2012, 2015), where better rates than $\frac{1}{\sqrt{n}}$ were obtained if some additional conditions are satisfied. Following the results from Rejchel (2015) we construct generalization bounds for boosting algorithms that are widely used in practice.

The chapter is organized as follows. In the next section we describe conditions that are sufficient to obtain generalization bounds for ranking estimators. Moreover, we show that they are satisfied for boosting algorithms. In Section 7.3 we study the quality of estimators on simulated data sets. The chapter is concluded in Section 7.4.

7.2 Assumptions, Main Theorem, and Application

There are two key assumptions needed to prove generalization bounds for ranking estimators. The first one is called the entropy condition and is a restriction on the capacity of a family of ranking rules that the estimator comes from. The second one states that the loss functions that we work with should be "strictly convex" in the sense that is given below.

We start by describing the entropy condition. The capacity of the class $\mathcal{F}$ is measured using its covering number. This approach can often be found in the literature (Pollard, 1984; van der Vaart and Wellner, 1996). Suppose that the family $\mathcal{F}$ is uniformly bounded by the constant A. In addition, let μ be a probability measure on $\mathcal{X} \times \mathcal{X}$ and let ρ_μ be a $\mathbb{L}^2$-pseudometric on $\mathcal{F}$ defined as

$$\rho_\mu(f_1, f_2) = \frac{1}{A}\sqrt{\int_{\mathcal{X}\times\mathcal{X}} [f_1(x_1, x_2) - f_2(x_1, x_2)]^2 d\mu(x_1, x_2)},$$

The covering number $N(t, \mathcal{F}, \rho_\mu)$ of the class $\mathcal{F}$ with a pseudometric ρ_μ and a radius $t > 0$ is the minimal number of balls (with respect to ρ_μ) with centers in $\mathcal{F}$ and radii t needed to cover $\mathcal{F}$. Thus, $N(t, \mathcal{F}, \rho_\mu)$ is the minimal number m with the property

$$\exists_{\bar{\mathcal{F}} \subset \mathcal{F},\, |\bar{\mathcal{F}}|=m} \quad \forall_{f \in \mathcal{F}} \quad \exists_{\bar{f} \in \bar{\mathcal{F}}} \quad \rho_\mu(f, \bar{f}) \le t.$$

In this chapter we are interested in families satisfying the entropy condition, i.e. there exist constants $D > 0$, $V \in (0, 2)$ such that for every probability measure μ on $\mathcal{X} \times \mathcal{X}$ and each $t \in (0, 1]$ we have

$$\ln N(t, \mathcal{F}, \rho_\mu) \le Dt^{-V}. \tag{7.3}$$

We have already mentioned that the second requirement that we need is related to the loss function ψ. To describe it we introduce the concept of the modulus of convexity (Mendelson, 2002; Bartlett et al., 2006).

Definition 7.1 The modulus of convexity of ψ is the function $\delta : [0, \infty) \to [0, \infty]$ defined as

$$\delta(t) = \inf \left\{ \frac{\psi(u_1) + \psi(u_2)}{2} - \psi\left(\frac{u_1 + u_2}{2}\right) : |u_1 - u_2| \ge t \right\}.$$

Now we can state the theorem that describes the quality of ranking estimators that are obtained by minimization of the empirical risk with the convex loss function. This originates from Rejchel (2015, Theorem 7.1).

Theorem 7.1 Let the family of ranking rules $\mathcal{F}$ satisfy the entropy condition (7.3) and be convex. In addition, we suppose that the modulus of convexity of the loss function ψ satisfies

$$\delta(t) \ge A_1 t^2$$

for some constant A_1 and each $t \in [-A, A]$. Then for every $x > 0$ with probability at least $1 - \exp(-x)$

$$Q(f_n) - Q(f^*) \le C_1 \max\left[\frac{1}{n^\beta}, \frac{\ln n}{n}\right] + C_2\, \frac{x}{n},$$

where $\frac{1}{2} < \beta = \frac{2}{2+V} < 1$ and C_1, C_2 are constants that depend only on the complexity of the model (i.e. values A, A_1, D, V), but they do not depend on the sample size n.

In Theorem 7.1 we state that the excess risk of ranking estimators can be upper bounded by the expression that decreases faster than $\frac{1}{\sqrt{n}}$. To obtain this result we need two non-restrictive conditions that relate to the convex loss ψ and the family $\mathcal{F}$. In the following example we show that they are satisfied for the ranking version of the AdaBoost algorithm. The proof of Theorem 7.1 can be found in Rejchel (2015).

Example 7.1 Let $\mathcal{R} = \{r : \mathcal{X} \times \mathcal{X} \to \{-1, 1\}\}$ be a family of "base" ranking rules with the finite Vapnik–Chervonenkis dimension $p > 1$. The output of the AdaBoost algorithm is an element of a convex hull of $\mathcal{R}$, namely

$$\mathcal{F} = \{f(x_1, x_2) = \sum_{j=1}^{T} w_j r_j(x_1, x_2) : T \in \mathbb{N}, \sum_{j=1}^{T} w_j = A,$$
$$w_j \ge 0,\ r_j \in \mathcal{R} \quad \text{for } j = 1, \dots, T\}.$$

From van der Vaart and Wellner (1996, Theorem 2.6.9) we obtain that for every probability measure μ on $\mathcal{X} \times \mathcal{X}$ and each $t \in (0, 1]$ we have

$$\ln N(t, \mathcal{F}, \rho_\mu) \leq Dt^{-\frac{2p}{p+2}}$$

for some positive constant D, so $\mathcal{F}$ satisfies the entropy condition with $V = \frac{2p}{p+2}$. Furthermore, the class $\mathcal{F}$ is obviously convex and the modulus of convexity of $\psi(x) = \exp(-x)$ fulfills on the interval $[-A, A]$ the condition $\delta(t) > \frac{t^2}{8\exp(A)}$. Thus, for the AdaBoost algorithm we obtain generalization bounds for the excess risk of f_n with the rate $\max\left(\frac{1}{n^\beta}, \frac{\ln n}{n}\right)$ for $\beta = \frac{p+2}{2(p+1)}$.

7.3 Experiments

In this section we study the quality of ranking estimators on simulated data sets. Estimators are obtained using the AdaBoost procedure, whose theoretical properties were investigated in the previous section. We use this algorithm in the version implemented in the "adabag" package (Alfaro et al., 2013) in R software (R Development Core Team, 2017). Briefly, AdaBoost is a "voting scheme" that works with a base family of ranking rules. This class usually contains simple decision trees such as decision stumps. The procedure works iteratively, namely at each step we train a ranking rule on the basis of a new sample that is generated from the initial one. The final estimator is a weighted average of decisions made by ranking rules calculated at each iteration. Even if the base class contains "weak" ranking rules, the final function performs remarkably well.

In experiments we assume that the relation between observations and variables determining the ordering between objects can be twofold:

$$Y_i = (\theta^*)^T X_i + \varepsilon_i, \quad i = 1, \ldots, n, \tag{7.4}$$

or

$$Y_i = e^{(\theta^*)^T X_i} + \varepsilon_i, \quad i = 1, \ldots, n, \tag{7.5}$$

where θ^* is the parameter, X_i is the observation, and ε_i is the noise variable. In addition, the number d of observations is 15 and the parameter θ^* is $(1, 2, 1, 2, 1, 2, \ldots)$.

We study three models: in model 1 (denoted by M1) we consider the linear relation (7.4) and $\varepsilon \sim N(0, 1)$. In model 2 (M2) we have errors with t-distribution and five degrees of freedom. In model 3 (M3) the nonlinear relation (7.5) and errors with standard normal distribution are considered. In each model observations are assumed to be independent and normally distributed.

For each model we generate a training sample $(X_1, Y_1), \ldots, (X_n, Y_n)$ for two sample sizes: $n = 50$ and $n = 100$. Next, for every sample we calculate the AdaBoost estimator. The quality of the estimator is evaluated on the newly generated test set, namely we take pairs of distinct objects from this set and check whether the ordering predicted by the estimator is the same as the true ordering between these objects. We repeat the experiment 30 times and the average proportions of the wrong predictions of the ordering are presented in Table 7.1. Standard deviations are given in parentheses. In addition, we can observe how the number of iterations influences the errors of estimators.

In our experiments the base class for the AdaBoost algorithm is the family of decision stumps. They are rather poor ranking rules that can be found in Table 7.1 in the

Table 7.1 Errors of estimators.

	$T = 1$	$T = 50$	$T = 100$	$T = 150$	$T = 200$
$M1$					
$n = 50$	0.403 (0.029)	0.174 (0.017)	0.143 (0.020)	0.136 (0.023)	0.128 (0.018)
$n = 100$	0.395 (0.029)	0.146 (0.011)	0.113 (0.007)	0.102 (0.010)	0.097 (0.010)
$M2$					
$n = 50$	0.411 (0.035)	0.212 (0.019)	0.194 (0.021)	0.192 (0.018)	0.186 (0.020)
$n = 100$	0.409 (0.025)	0.176 (0.020)	0.154 (0.018)	0.148 (0.021)	0.146 (0.022)
$M3$					
$n = 50$	0.425 (0.021)	0.253 (0.028)	0.241 (0.030)	0.238 (0.030)	0.235 (0.029)
$n = 100$	0.411 (0.025)	0.231 (0.018)	0.222 (0.022)	0.215 (0.019)	0.214 (0.019)

$T = 1$ column (i.e., only one iteration is made by the algorithm). Their errors are about 40%, so they behave in a similar way to random guessing. However, if we combine these weak ranking rules, then their errors decrease (the improvement is significant even for $T = 50$). The results for $T = 150$ and $T = 200$ seem to be similar, so we could say that the errors of our procedures "stabilize". Obviously, looking at results for $n = 50$ and $n = 100$ we can see that greater sample sizes lead to smaller errors.

Comparing results for different models we can state that in the first model the relation between observations X_i and variables Y_i is linear, and the procedural errors are the smallest. The second model is more difficult to work with because outliers appear. However, algorithms can overcome this quite well and their quality is only about 5% less. Predicting the ordering between objects in the third model seems to be relatively hard. However, errors of procedures are between 20% and 25% (except $T = 1$), so they work pretty well even in this case.

7.4 Conclusions

We have considered the ranking problem that relates to predicting the ordering between objects on the basis of their observed features. In the first part of the chapter we investigated the theoretical properties of ranking estimators that are obtained by convex empirical risk minimization. In the second part we studied the quality of estimators in predicting the ordering between objects on simulated data sets. The results we obtained confirm the usefulness and high accuracy of the estimators considered.

8

Classification and Regression Trees for Analyzing Irrigation Decisions

S. Andriyas and M. McKee

8.1 Introduction

In this data-rich world, there is a lack of pertinent information about certain phenomena that are either hard to model or lack a complete physically based cause–effect description of the problem. This presents challenges in the use of conventional approaches such as deterministic models to predict future conditions. Such a problem exists in understanding and predicting a farmer's decision to irrigate. Substantial scientific theory and large quantities of data are available to analyze the irrigation problem and forecast short-term irrigation demand, but the problem of accurately anticipating the short-term water demand of an individual irrigator still remains. This is due to a limited understanding of the irrigation practices that are followed by different farmers and how farmer preferences influence decisions about the timing of irrigation.

The site selected for this study is equipped with technologies to monitor reservoir releases and canal diversions, and it has dependable forecasts of evapotranspiration (ET). Some of the irrigated fields have real-time soil moisture measurements to study agricultural water use in the area. In spite of these developments, day-to-day irrigation demands are difficult to forecast. Information about such demands can be vital to help irrigation system operators achieve greater efficiency in water deliveries. In an on-demand irrigation delivery system, farmers make the basic water use decisions. Hence it is essential to consider their decision-making mechanisms in forecasting short-term irrigation demand.

Irrigation behavior has rarely been a topic of research. Each farmer has personal goals to achieve in a season, ranging anywhere from profit maximization, to crop quality, to being environmentally conscious about saving water. A farmer whose primary profession is agriculture will make different choices from the one who considers agriculture as a secondary occupation. These characteristics make it more difficult to forecast behavior. The few studies that have dealt with irrigation behavior have been inconclusive in understanding the factors that contribute to decisions regarding if and when to irrigate. To find out the scope of studies done on farmer's behavior previously, we present some of the notable ones in the field.

Becu et al. (2006) used a multi-agent system for a study of water sharing between two villages located at the upstream and downstream ends of a watershed. The objective was to evaluate various options to allot water to the villages and provide feasible

Ensemble Classification Methods with Applications in R, First Edition.
Edited by Esteban Alfaro, Matías Gámez, and Noelia García.

solutions to different water users for dealing with water scarcity. The solutions were found by analyzing the impact of different land use and water management options on water deficit. Since it involved water use decisions, farmers' behavior was considered in terms of what crops are planted, when they are harvested, and how they are irrigated during the season. Farmers were grouped into various classes on the basis of different cropping patterns identified in the region.

This study simulated irrigation decisions taken by the farmers on the basis of the crops they were growing. The paper primarily evaluated the use of a multi-agent system to support collective decision-making in a participatory modeling framework. The farmers initially had misunderstood the model as being a representation of the real world but the model tested scenarios and suggested possible solutions. The study showed that the players involved from the upstream village were concerned about the impact of water scarcity on both villages, while the ones downstream were only locally concerned. Bontemps and Couture (2002) developed a sequential decision model to study water use behavior under conditions when the farmers paid a negligible amount to obtain water and there was no charge for supplying it. The model required precisely calibrated crop-growth simulation models, irrigation practices, and information about land use and climate for the region of interest. Data were obtained by integrating an agronomic model, and a solution-searching optimization methodology connected to an economic model. Irrigation demand functions were estimated using non-parametric methods. Three different functions were estimated keeping three types of representative weather conditions in mind. The method was applied to estimate crop demands in southwest France. Results for all types of weather conditions were found to be same (demand function curves had the same shape: decreasing and non-linear), suggesting that irrigation demand was inelastic for the small amount of water available, but if the total quantity of water was increased, the demand became more elastic.

The results showed that the threshold price at which alteration of price-response seems to take place depends on weather conditions. Le Bars et al. (2005) developed a multi-agent system paradigm to simulate farmer-agents and their decisions over a number of years under conditions where water supply was limited. The water manager controlled the amount of water given to a farmer by using allocation rules that were based on the amount of water requested by farmers at the beginning of the season. The farmer-agents each owned a farm with several plots and could decide their own cropping plan. Weather variables were random. This agent-based model helped the negotiations between water suppliers, farmers, public servants, and environmentalists by presenting the impacts of water-allocating rules based on different criteria. In other words, rules can be tested and resulting consequences can be seen. It was found that for global corn profits, based on the information of whether the previous agent knows about the water allocated to the agent before them, the differences between farmers could be decreased. This decrease would also show a drastic effect on water use efficiency.

From the limited literature on farmer irrigation decision behavior, it is clear that few studies have been conducted to analyze decisions already made or to forecast future irrigation decisions under simulated conditions. Models that could provide such forecasts could be potentially useful for improving irrigation system operations. The study reported here is a first attempt at analyzing farmers' decisions using "decision" trees. We use data about the biophysical conditions during the growing season to isolate information available to the farmer about differences on the days leading up to the time of irrigation. We also look into the possibility of using those differences to forecast farmer decision-making.

8.2 Theory

A wide range of machine learning techniques is available today to address modeling problems where missing information is an issue. These show promise for the analysis of problems involving the forecasting of decision behavior under conditions where it is not possible to quantify all of the process-specific factors that affect the decision. Figure 8.1 shows a tree structure. The nodes are the variables related to the process in the form of root, intermediate or terminal nodes (which do not have any child nodes). As we descend in the tree the importance of the variable to the process decreases. The variable at the root node is the most important. The effect of all the variables leading to the terminal node is collective.

Trees are used to understand systems that have little a priori information about how and which variables are related. Classification trees have been used by Kastellec (2010) to analyze judicial decisions and laws. These are the applications of trees to real problems.

Decision trees are a powerful tool for classification and forecasting. The features and capabilities of the trees are described in Hill and Lewicki (2007). They can give insights into non-parametric, non-linear relationships between a large number of continuous and/or categorical predictor inputs and output variables, which may be continuous or categorical. When the output variable is continuous it is a regression analysis and when categorical it is a classification problem. Decision trees divide a heterogeneous group of features into small homogeneous groups with respect to the output variable. A binary tree formed by two child nodes split from each parent node is one such structure. The split is best when it separates the data into groups with two different predominant classes. "Levels" in a tree are referred to as the depth of the tree, and "size" is the number of nodes in the tree. The measure often used to estimate the split is known as "purity". The best split is defined as the one which increases the purity of subgroups by a considerable amount and creates nodes of similar size (not very small ones). Dense structures can often be simplified by pruning. The tree models make no prior assumptions about the data. No unit conversions are required. Raw data can be used as it is. The variables at the root of the tree are deemed the most important. For a more detailed description of classification trees, see Section 1.5.

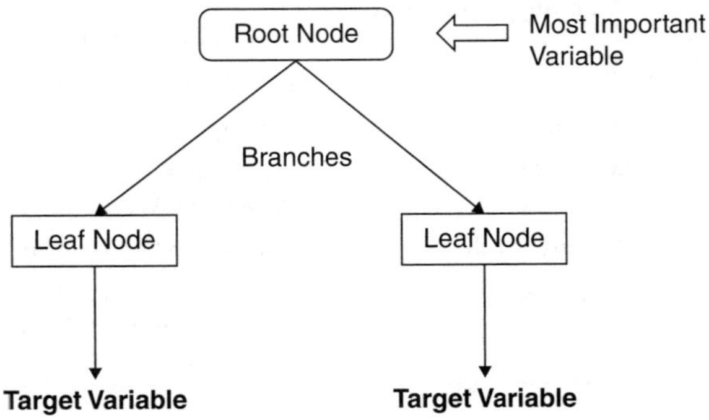

Figure 8.1 A tree structure.

8.3 Case Study and Methods

8.3.1 Study Site and Data Available

The data used in this study are from the Canal B region of the Lower Sevier River Basin, near the town of Delta in south-central Utah. This area covers approximately 20 square miles of irrigated farm land. Alfalfa, barley, and corn are the main crops grown in the area. Irrigation consumes a large amount of water in this basin. Weather data for Delta was obtained from the following website: http://www.cemp.dri.edu/cgi-bin/cemp_stations.pl?stn=delu. Data estimated using Kimberly Penman Reference ET rules are also available on this website.

Table 8.1 presents the variables used to build trees and predict the irrigation decision. Variables 1, 4, 5, 6, and 21 are weather variables. Data were also available for canal flow

Table 8.1 Predictor variables, the represented factors as seen by the farmer, and the target variable used for trees analysis.

Sample number	Variable name	Represented factor	Continuous OR categorical (no. of classes)
1	AirTemp	Average air temperature	Continuous
2	GrowingDegDays	Growing degree days accumulated till a given day and reset on the day of irrigation	Continuous
3	GrowStageIrrigNeed	Sensitivity of growth stage to water stress as indicated by growing degree days	Categorical (2)
4	WindSpeed	Wind speed	Continuous
5	RH	Relative humidity	Continuous
6	ET	Potential evapotranspiration (ET)	Continuous
7	ET_c	Crop rvapotranspiration	Continuous
8	CropCoeff	Crop-specific coefficient	Continuous
9	SoilStressCoeff	Soil stress coefficient	Continuous
10	ET_a	Actual evapotranspiration	Continuous
11	$CumET_c$	Cumulative crop ET	Continuous
12	StressIrrigNeed	Consumptive use as indicated by $CumET_c$	Categorical (2)
13	CanalFlow	Canal flow rates	Continuous
14	WaterSupplyIrrigNeed	If the farmer irrigated when his neighbors irrigate as indicated by CanalFlow	Categorical (2)
15	JDay	Julian day in the season	Continuous
16	WeekEndORNOT	Saturday/Sunday	Continuous
17	WkEndIrrigNeed	If the farmer irrigated on a weekend as indicated by WeekEndORNOT	Categorical (2)
18	RootingDepth	Rooting depth of the plant	Continuous
19	CropIrrigNeed	Plant need indicator: the deeper the root, the more frequent is the need for water as indicated by RootingDepth	Categorical (2)

(*continued*)

Table 8.1 (Continued)

Sample number	Variable name	Represented factor	Continuous OR categorical (no. of classes)
20	SMCinit	Soil moisture content at the start of the day	Continuous
21	Rain	Precipitation amount	Continuous
22	AmountPercolation	Amount of irrigation water percolated	Continuous
23	IrrigationAmt	Estimated amount of irrigation from the soil moisture probes	Continuous
24	DepInit	Depletion at the start of the day	Continuous
25	DepEnd	Depletion at the end of the day	Continuous
26	SoilIrrigNeed	If the soil is dry or not (also indicated in plant condition) as indicated by SoilStressCoeff	Categorical (2)
27	Year	Year, indicating a dry, moderate or wet year	Categorical (4)
28	Yield	Yield estimated using ETa and ETc	Continuous
29	MarketPrice	Price of the crop	Continuous
30	ProfitORLoss	Profit or loss for the farmer	Continuous
31	EconIrrigNeed	Economic need to irrigate the crop as indicated by ProfitORLoss	Categorical (2)
32	ID	Different farmers	Categorical (39)
33	SoilType	Type of soil	Categorical (2)
34	Irrigate	The irrigation decision	Categorical (2), (0–1 for regression and Yes–No for classification)

rate (Variable 13). Three types of soil are found at Delta: silty clay loam, silty clay, and loam. Farmer identification numbers convey information to the model that the data are from a different subject. Soil moisture contents were available from numerous soil moisture probes installed in the Canal B irrigation command area, and were corrected using a mass balance constraint on soil moisture. Variables 7–10, 18, and 20–25 are required for the soil moisture balance computation. A daily time series was created for market prices of alfalfa, corn, and barley using the monthly data available from the USDA website for Millard County, Utah. Approximate planting dates were established by initiating the soil moisture calculations from a random day such that soil moisture matched the day of first irrigation, which was known from the soil moisture probe data. We assumed that the initial depletion was zero and began the computations from field capacity.

Phenology coefficients (K_c) for all the crops were derived from Wright (1982) and FAO-56 (Allen et al., 1998), and were found to be quite representative. Since we were using the values of K_c for crop reference ET, we had to multiply the values by a factor of 1.2 to model a field crop instead of grass reference ET. All the other variables were either derived from the primary data or categorized to simplify their representation in the model. If both the data and the derived variables behave the same, then the derived ones can be removed.

8.3.2 Model, Specifications, and Performance Evaluation

The models used in this study were implemented using R-statistical software (R Development Core Team, 2017). For all the classification problems it is necessary that the target variable be categorical (e.g. "Yes" and "No"), which can be done using the *factor* function. The data were randomly partitioned into training and testing sets. For all the data sets, one-fourth of the data were used for testing. The input to all these algorithms for our case was the decision to irrigate ("Yes") or not ("No"). During training, the model was tuned according to the irrigation decision. During testing, this target variable was forecasted. The outputs of the model were the confusion matrix and the error rate. We used the accuracy rate (calculated as the difference between 100 and the error rate) and the confusion matrix to evaluate model performance.

To do a CART analysis, we used the **rpart** (Therneau and Atkinson, 2017) package in R. It is powerful and easy to use, and is based on the same algorithm as Breiman et al. (1984). During the model fitting process, the *rpart* function was applied to the training data with the dependent variable being the irrigation decision and method = "class". The *predict* function uses the fitted tree and predicts the classification for the test data set. The *table* function can be used to obtain a confusion matrix, and the accuracy rate can also be obtained. The accuracy rate is the sum of the diagonal elements of the confusion matrix, divided by the sum of all the elements. The *printcp* function can be used to print the complexity parameter (CP) table (Table 8.2 shows a sample output) for the fitted tree. From this we can find the optimal pruning of the tree based on the CP. It can be seen in Table 8.2 that nsplit (number of splits) denotes the size of the tree and (nsplit + 1) is the number of nodes in a tree. Scaled errors are presented such that the error at the first node is 1. Using the 1-SE rule to find the best number of splits, the smallest "xerror" is added to the corresponding "xstd", as shown in the last column. The number of splits resulting in the smallest error is the "best split", e.g. in this case the optimal number of splits is 1. For this case the pruned tree will have two nodes. If in the case where the sums of "xerror" and "xstd" are all equal, the "best split" will be the tree with the fewest number of splits other than zero, which would leave only the root node. To assess the performance of a predictive model, we have also used cross-validation. We cross-validated using bagging (Breiman, 1996a), derived from bootstrap aggregating. Bagging is an ensemble method that decreases the variance of the original individual models by using a bootstrap of the training set to build every new model and then taking the average of the predictions from

Table 8.2 The cost-complexity parameter (CP), relative error, cross-validation error (xerror), and cross-validation standard deviation (xstd) for trees with nsplit from 0 to 8.

Sample number	CP	nsplit	Relative error	xerror	xstd	Sum of xerror and xstd
1	0.709	0	1.000	1.069	0.050	1.119
2	0.030	1	0.291	0.291	0.035	0.326
3	0.020	5	0.172	0.345	0.038	0.382
4	0.015	7	0.133	0.330	0.037	0.367
5	0.010	8	0.118	0.310	0.036	0.346

those models. For running 10-fold cross-validation with bagging we used the **adabag** (Alfaro et al., 2013) package in R. For cross-validation, two-thirds of the data were used for training and the remaining for testing. The function "bagging.cv" requires the target variable as input from the training data set and the number of iterations, "mfinal" variable (default =100). The most important variable is at the root node.

8.4 Results and Discussion

Figure 8.2 presents a plot of some of the weather variables to illustrate some existing groups based on the irrigation decision. No obvious classes can be found in the plot.

If apparent groups existed, we would not use such algorithms. Usually in practical problems it is difficult to find classes to discern groups based on the target variable. We have to then seek the help of specialized techniques like recursive partitioning, in our case, for exploring the groups, if any.

Since the day of irrigation can be anywhere from 24 hours to 3 days from the day of order (call-time), we decided to examine the importance of the variables presented in Table 8.1 for:

- all the days in the season (referred to as 'all days" in the results)
- four days before the day of irrigation and the day itself (referred to as "4-day" in the results)
- the day of irrigation and a day before (referred to as "1-day" in the results).

The prediction accuracy for the tree analysis is presented in Table 8.3. The best results are displayed in bold. It can be seen that the predictions of the models, which were given the description of the whole season (i.e., the all-days model), performed better than the 1-day or 4-day models. Given that there is a possibility of some missing information, the algorithm worked exceptionally well to predict future decisions. CART had accuracy estimates of 99.2%, 98.7 and 96.9% for alfalfa, barley and corn, respectively.

A 10-fold cross-validation was done on all the data sets and the cross-validation accuracy was reasonably close to the resubstitution accuracy, except for the corn 1-day model, as shown in Table 8.3. Cross-validation is performed to help determine if the classifier is being overfitted. For our case, the performance is promising. Figure 8.3 shows trees built for the three crops in the study. Clearly there are different strategies for the three crops, starting with the same variable. The most important variable for all three crops was cumulative crop ET, or the consumptive use.

According to many standard agricultural handbooks, crop growth stage and its sensitivity gauged by accumulated heat units (or growing degree days, GDD), and the consumptive use that this implies, is almost always considered by farmers when deciding about irrigation. StressIrrigNeed is the indicator for critical $CumET_c$, so it was expected that it would be next in importance in explaining irrigation decisions. At each terminal node, there is a group based on a variable or its level which makes it different from the others, according to CART interpretation. In other words, there is a class at every node where we see either a "'Yes" or a "No" response. Since we are interested in the reasons behind irrigation decisions, we are only evaluating "Yes" responses.

Figure 8.3 shows that the alfalfa growers in the first group were irrigating when their neighbors irrigated (canal flow > 209.5 cfs, which was a high flow rate), at medium

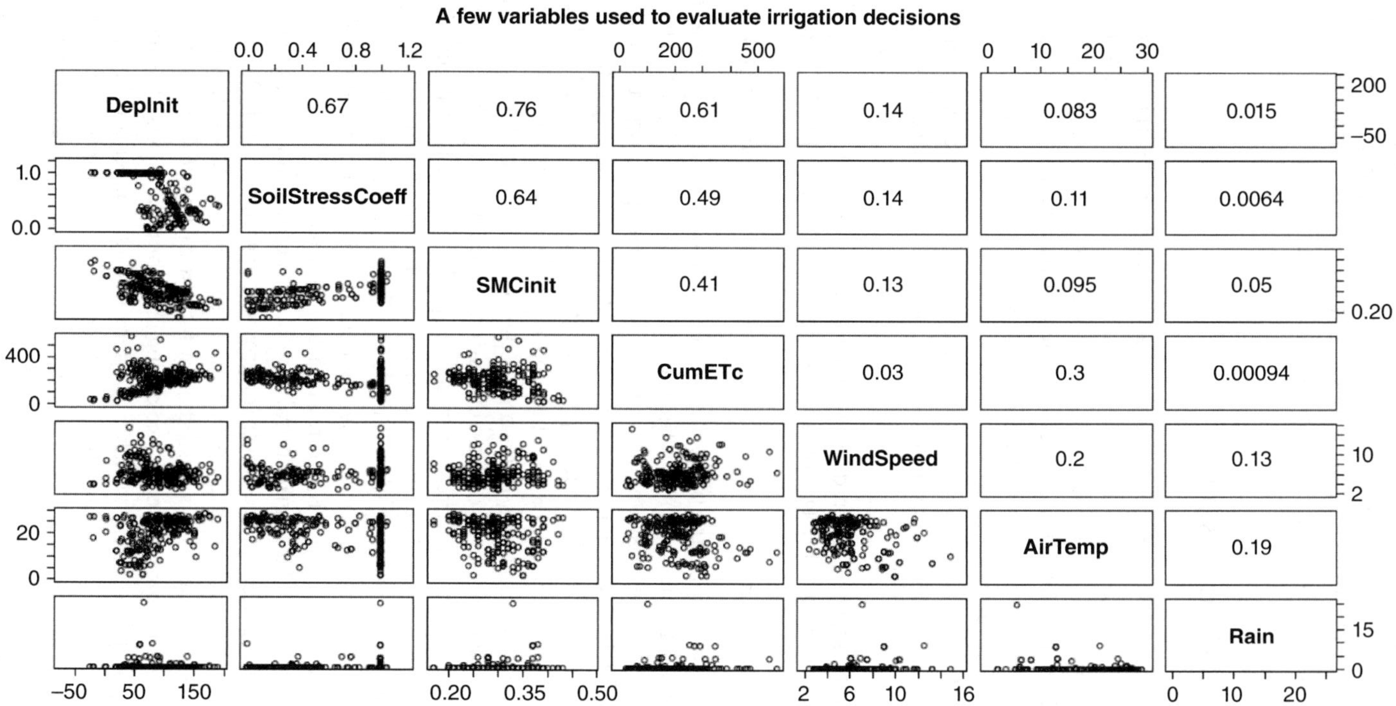

Figure 8.2 Pairs plot of some weather variables used in the tree analysis, with the intention of finding groups of similar features.

Table 8.3 Accuracy estimates on test data for CART models. Resub, resubstitution accuracy estimate; Xval, 10-fold cross-validation accuracy estimate. (a) 1-day, (b) 4-day, (c) all days models.

Crop	Alfalfa			Barley			Corn		
Model	(a)	(b)	(c)	(a)	(b)	(c)	(a)	(b)	(c)
CART-Resub	83.5	91.1	99.2	48.5	83.1	98.7	66.0	84.9	96.8
CART-Xval	83.3	94.5	99.4	53.4	81.2	98.0	46.0	82.1	97.1

soil moisture depletion (> 65.16 mm), and when farmers other than the ones shown in Figure 8.3 were irrigating according to these common rules. Since we wanted to avoid bias in selecting the training and testing sets, we used bootstrapping to sample the data sets. It is important to note that these data sets are a mix of information from any of the four years (2007–2010). The second class evidently used the $CumET_c$ (StressIrrigNeed = Yes) measure to time the irrigations. $CumET_c$ is the crop ET accumulated between irrigations and is the same as depletion. As a general rule, alfalfa is primarily irrigated either before or after the cuts. All the resulting principles conform to recommended irrigation practices, since depletion would not be used generally as one of the indicators to trigger irrigation for alfalfa. The possibility that farmers are choosing to irrigate when their neighbors are irrigating is also refuted since the fields involved have different crop planting dates. This hints at different maturity and cut timings, and might even point at a different crop quality.

Barley irrigation strategies (Figure 8.4) were very straightforward, with the low depletion of 22 mm and consumptive use (>170 mm) being the only indicators CART could discriminate. This suggests that the farmers who were growing barley were not taking risks with respect to the irrigation because this would mean loss in yield to them, even though they could wait longer to irrigate. We did not have enough data to draw strong conclusions but these may be probable reasons for the observed timing of irrigation.

For corn (Figure 8.5), CART presented a huge tree with many variables. It clearly showed three classes (i.e., three terminal nodes with "Yes" as the irrigate decision). For the first group, the day in the growing season (JDays between 127 and 145) were crucial and an irrigation amount more than 105 mm appeared to be the driving factor. The day in the season is indicative of a certain critical growth stage for corn. The irrigation amount may seem a strange choice for grouping farmers, but it implies that farmers who replenished the soil moisture to this level would irrigate similarly. This corresponds indirectly to consumptive use. For the second group, in years 2007 and 2010, the farmers other than those shown irrigated while ET was constraining on the crop. This meant that temperatures could have been high for long periods when the farmers decided to irrigate. The third group irrigated when consumptive use was more than 122 mm and the predicted market price was higher than before. The consumptive use for corn is always a driving force for irrigation. If there is high moisture stress, the amount of carbohydrates available for kernel development in corn is inhibited, which can affect the yield. The implication is that corn growers were keeping the stress levels in control by irrigating at moderate levels of moisture depletion.

CART pruned trees performed the same as the fully grown versions, but the advantage was the smaller number of variables for interpretation. We have opted for the fully

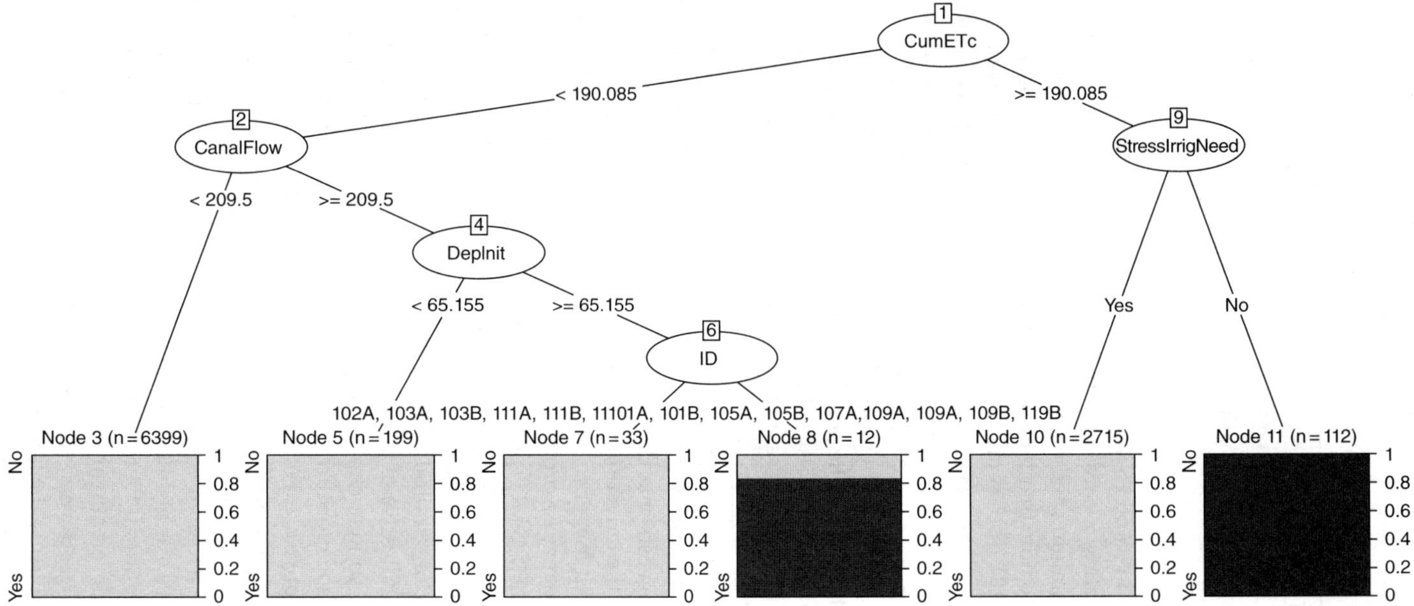

Figure 8.3 CART structures for alfalfa decisions.

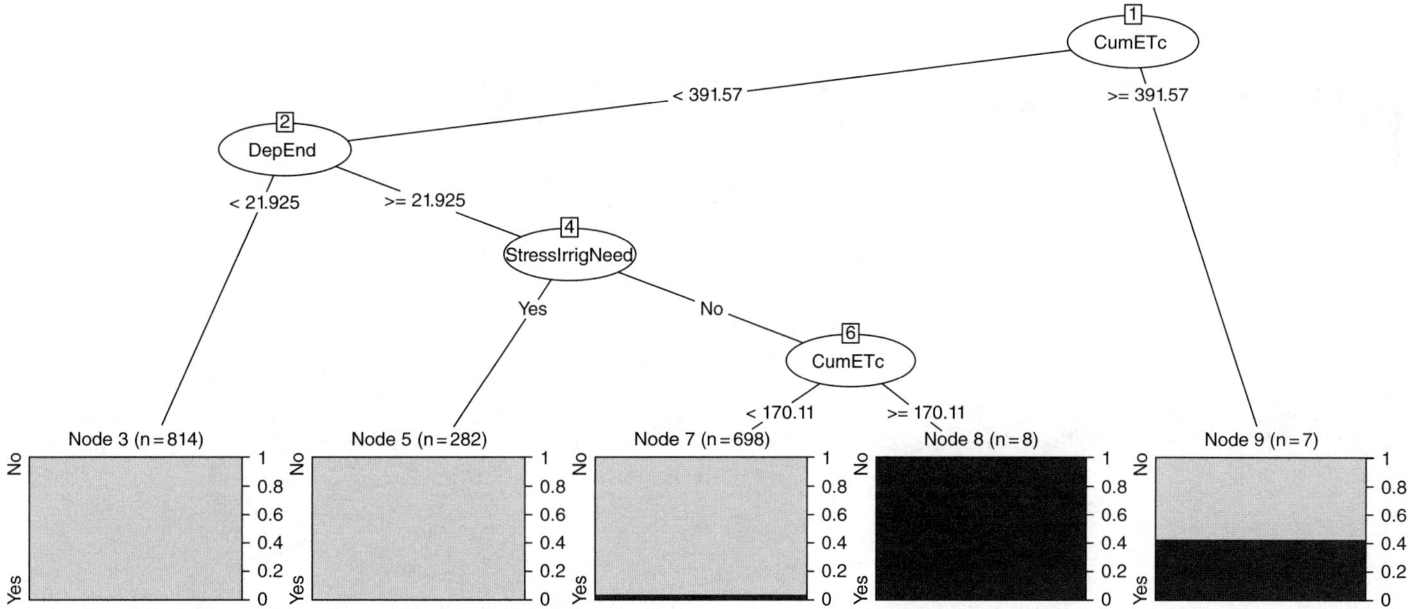

Figure 8.4 CART structures for barley irrigation decisions.

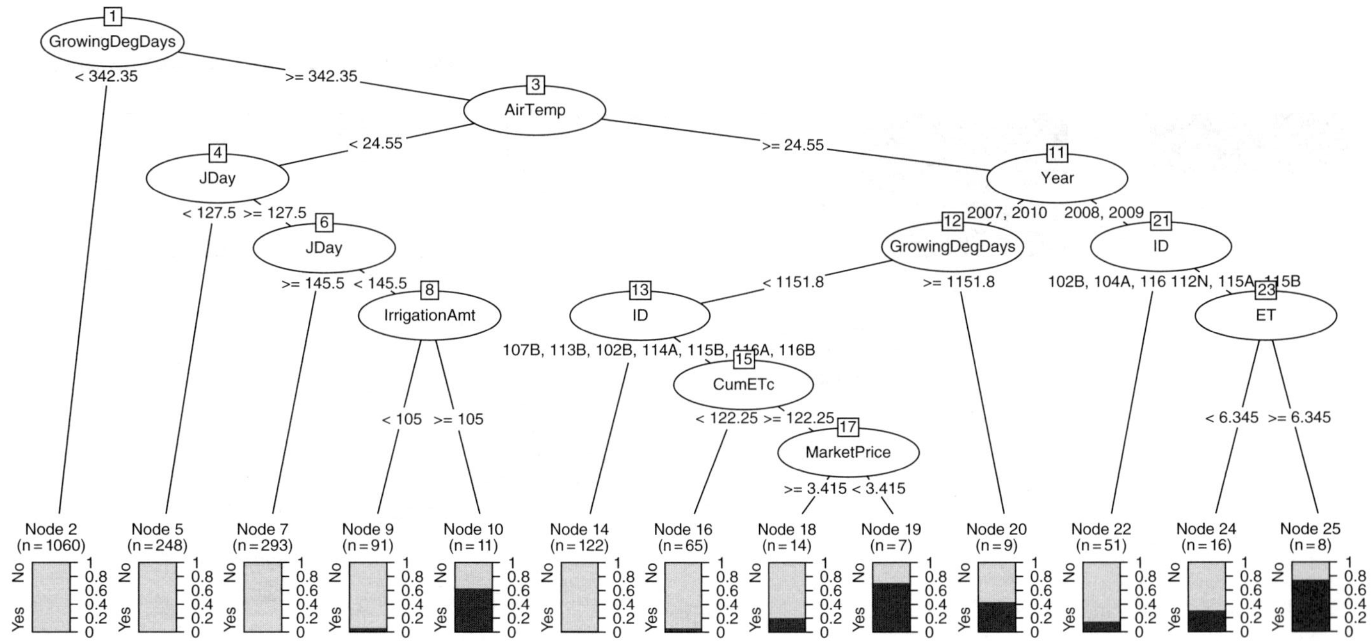

Figure 8.5 CART structures for corn irrigation decisions.

grown versions, however, since it gives us an in-depth analysis of the factors leading to irrigation. Though the pruned tree narrowed the choices of variables it does not refute the fact that farmers consider multiple factors in the thought process of scheduling irrigation.

8.5 Conclusions

Irrigation system managers would benefit from information about short-term irrigation demand. This study applied different types of classification trees to infer how farmers, the water users, make irrigation decisions. This information can be used to predict future actions and forecast short-term water demands, relying on readily measurable biophysical data alone as input. The results from this study show that biophysical conditions can be used as indicators of irrigation behavior and have the potential to be used as predictors for future irrigation decisions.

The tree algorithm provides analysis of the factors leading to decisions and presents a possible forecasting tool. CART is a classification algorithm and presents simplified trees. In terms of modeling different problems, it is important to tune the models and find the best-fit parameters to improve accuracy estimates. It was found that the model had high classification accuracy to predict irrigation decisions when larger data sets (more information) were used. Smaller data sets supplied incomplete information to the models, resulting in poor classification rates.

The models picked logical factors which can possibly lead to an irrigation decision. From the point of view of the crops to be studied, model performance cannot be compared for perennial or annual crops since the database for alfalfa was bigger than that for the other two crops. Information for the whole growing season can increase the number of cases for training, resulting in better model performance.

Table 8.4 summarizes the probable important factors exhibited in the tree structures and variable importance measures. The predictors which are most useful in forecasting irrigation decisions are consumptive use, growing degree days or cumulative temperatures, and irrigating when a neighbor irrigates. The variable Year is specific to a perennial crop like alfalfa. Since ET is dependent on temperature, temperature and canal diversion measurements can be used to forecast farmers' future actions. The other important aspect in getting accurate forecasts is the amount of information given to the model. Information for the full growing season should be provided, which means that the models will not be able to handle missing information for this problem. This

Table 8.4 Important variables for irrigating different crops according to CART.

Alfalfa	Barley	Corn
$CumET_c$	$CumET_c$	GDD
Canal flow	Depletion	AirTemp
		Day
		Year

feature is similar to a farmer managing his farm who monitors day-to-day crop and soil conditions and makes decisions accordingly. If he skips a few days in observing these conditions, he will not be able to make appropriate decisions due to the gap in information. We conclude that the most important factor for irrigation behavior appears to be crop need, followed by farmers' observations of their neighbors' actions. These findings are promising and can be used to make estimates of short-term demand forecasts.

9

Boosted Rule Learner and its Properties

M. Kubus

9.1 Introduction

With the rapid development of information technology, large databases are collected in many areas of human activity, for example supermarket transaction data, credit card usage records, and telephone call details. The use of information hidden in databases supports a decision-making process. To discover unknown and unexpected patterns in the data, a variety of data-mining techniques are applied. For several reasons, classification trees (Breiman et al., 1984; Quinlan, 1993) have gained great popularity. They can deal with various types of features as well as with missing values. As a non-parametric method, classification trees do not require any assumptions about distributions and they are robust to outliers. Moreover, using natural language a tree is easy to interpret, which enables an insight into the phenomenon of interest. Usually, real-world domains contain irrelevant variables that have no impact on the response variable. Such variables lead to overfitting and they can reduce the predictive accuracy of the model on unseen data. Classification trees belong to embedded methods of feature selection (Guyon et al., 2006). This method introduces into the model only such variables that locally optimize the homogeneity criterion, thus making the automatic feature selection. All these characteristics of classification trees make them a convenient tool that does not require any pre-processing steps. The drawback of this method is low stability, which means that small changes in the training sample affect the form of classifier and often its accuracy.

The improvement in stability and accuracy can be obtained by aggregating the classifications of a greater number of trees. This idea reflects the collective decision-making by the people, and has a strong theoretical justification. Condorcet Jury Theorem on group decision-making was formulated in 1785 (Condorcet, 1785). It has been shown that the error of the ensemble is lower than the mean error of the base models (Breiman, 1996b; Fumera and Roli, 2001). The research on bias and variance decomposition of the classifier has shown that an ensemble improves the stability of the model (Breiman, 1996b; Freund and Schapire, 1997). Dietterich (2000a) gives three theoretical arguments why an ensemble may work better than a single classifier. There are two main techniques of combining the base models in an ensemble. Breiman (1996a) proposed bagging, where trees are constructed on bootstrap samples of the original training set. Freund and Schapire (1997) introduced weighting of objects so that those which were most often misclassified by the former classifiers could be resampled with a higher

Ensemble Classification Methods with Applications in R, First Edition.
Edited by Esteban Alfaro, Matías Gámez, and Noelia García.

probability in the next iteration. Base models in an ensemble should be unstable, and their accuracy at least slightly better than the classification based on prior probabilities. The outputs of the base models are usually aggregated with a use of majority voting, although more sophisticated techniques are also possible (Górecki and Krzyśko, 2015).

From the early years of machine learning rule induction has been developing in parallel with classification trees, and it has much in common with them. The classifier is a set of logical rules in a form of implication. The antecedent is a conjunction of the simple conditions that take one of the following forms: $x_j = v$ for nominal variables (v is a category) or $x_j \geq l$ ($x_j \leq l$) for quantitative variables. The consequence of the implication is a decision about classification. Such representation is known as the disjunctive normal form (DNF), but other forms in rule induction are also possible (Fürnkranz, 1999). Note that the classification tree can also be viewed as a DNF. On the other hand, not every set of rules in DNF has a hierarchical tree structure. For this reason rule induction can potentially detect a wider range of patterns in the data. There are two popular approaches to building a set of rules. The first one uses a classification tree that is decomposed into a set of rules and then pruned to reduce the error (Quinlan, 1993). The second one, proposed for the first time by Michalski (1969), follows the separate-and-conquer algorithm, which iteratively generates the rules identifying the most homogeneous regions in the feature space. A review of different realizations of this general algorithm was presented by Fürnkranz (1999).

Analogously to classification trees, further development of rule induction methods focuses on the ensembles. Friedman and Popescu (2008) proposed decomposition of a bagged family of trees into the set of rules, which is then weighted by regularized linear regression. Pfahringer et al. (2004) proposed an approach similar to random forests (Breiman, 2001). They built an ensemble containing a large number of randomly generated rules. Cohen and Singer (1999) introduced boosting to the rule induction instead of separate steps. Their method is the subject of empirical studies in this work.

In this paper we examine some properties of the boosted rule learner. We ask whether boosting improves the stability of a classifier in this form, and whether this method is robust to irrelevant variables. We also compare the classification accuracy of two algorithms that have implemented boosting: AdaBoost.M1, which builds an ensemble of trees, and SLIPPER, which builds a set of weighted rules. In our experiments we use real data and artificially generated irrelevant variables.

9.2 Separate-and-Conquer

One of the most advanced algorithms that follows the separate-and-conquer scheme is RIPPER (Cohen, 1995). Its effectiveness is a result of a quite sophisticated pruning technic, which combines pre- and post-pruning, and a use of stopping criteria based on the minimum description length (MDL) principle (Rissanen, 1978). Consider a classification task where a training set is given:

$$\{(\mathbf{x}_1, y_1), \dots, (\mathbf{x}_m, y_m)\} \; : \; \mathbf{x}_i \in \mathbf{X} = (X_1, \dots, X_d),\; y_i \in Y, i \in \{1, \dots m\}. \tag{9.1}$$

The purpose is to estimate the model that would predict the class (Y is nominal) for unseen objects, which are supposed to appear in the future. For every class in turn the algorithm repeats two steps. After learning the single rule (conquer step), the algorithm removes from the training set the objects that satisfy this rule (separate step). This is continued until no objects from a chosen class remain or the algorithm reaches some stopping criteria. First, the training set (equation 9.1) is randomly split into a training sample and a test sample. The procedure of generating a single rule requires a choice of search strategy, evaluation function, and optionally stopping criterion. RIPPER performs a greedy search of the space of all possible simple conditions. They are added to the rule R to optimize the function:

$$f(R) = -p\left(\log\frac{p'}{p'+n'} - \log\frac{p}{p+n}\right) \tag{9.2}$$

where p and n are the numbers of objects covered by the rule R (p from the chosen class and n from the others), and p' and n' are the numbers of covered objects before adding the condition. Immediately after constructing a rule, it is pruned using a test sample. The final sequences of conditions are removed in a greedy manner to maximize the function:

$$f(R) = \frac{p-n}{p+n'} \tag{9.3}$$

where p and n are the numbers of covered objects from the test sample. In turn, generating the rules is stopped when introducing the next rule results an increase in description length of more than b bits from the smallest length so far obtained. Cohen (1995) suggests $b = 64$. After learning the rule set, it is pruned again. For every rule in turn, its two alternatives are constructed and finally one chooses this rule, which optimizes the MDL heuristic. Cohen (1995) has shown for the 37 benchmark problems that RIPPER is more efficient that C4.5 rules (Quinlan, 1993), especially on noisy data sets.

9.3 Boosting in Rule Induction

Cohen and Singer (1999) adapted boosting for inducing the rules in the SLIPPER algorithm. In a contradiction to the ensemble of trees, their application of boosting leads to a single and comprehensible model. Instead of removing covered objects in every iteration, their weights are reduced so that the other objects are covered in the next iteration with more probability. In this way, every rule is generated using an entire training set. The rules are also weighted, which gives the possibility of ranking their importance, which is desirable with respect to interpretation.

The SLIPPER algorithm was proposed for the binary classification problem. We are given a training set (equation 9.1) where the classes are coded with $\{-1, +1\}$. The weights of the objects are initialized with equal values $D(i) = 1/m$. The algorithm iteratively repeats two steps. The number of iterations can be set arbitrarily or via the cross-validation. In the first step, SLIPPER learns a single rule using the procedure from RIPPER (Cohen, 1995) and then, in the second step, the weights of the objects are updated. Thus, the covered objects are not removed from the training set, but they obtain weights according to the generalized version of AdaBoost (Schapire and Singer, 1999):

$$D_{t+1}(i) = \frac{D_t(i)e^{-y_i C_{R_t(i)}}}{Z_t} \tag{9.4}$$

where

$$C_{R_t} = \frac{1}{2} \ln \frac{W_+}{W_-} \tag{9.5}$$

is a weight of the rule R_t and

$$Z_t = W_0 + W_+ e^{-C_{R_t}} + W_- e^{C_{R_t}} \tag{9.6}$$

is a normalization factor where

$$W_0 = \sum_{x_i \notin R_t} D(i), \quad W_+ = \sum_{x_i \in R_t; y_i = +1} D(i), \quad W_- = \sum_{x_i \in R_t; y_i = -1} D(i). \tag{9.7}$$

Note that $\mathbf{x} \in R$ means that x satisfies rule R. When a rule covers only the objects from the class which it has been covering, W_- is equal to 0. Therefore equation 9.5 is modified as below:

$$\tilde{C}_{R_t} = \frac{1}{2} \ln \frac{W_+ + \frac{1}{2m}}{W_- + \frac{1}{2m}}. \tag{9.8}$$

Schapire and Singer (1999) showed that the resubstitution error is minimized when the weak learner algorithm (here RIPPER) picks a single rule and its weight to obtain the smallest value of normalization factor Z_t (equation 9.6). Hence, as a weight of a rule is given in equation 9.5, we can conclude that the function of the criterion in the RIPPER procedure of generating a single rule should take the form

$$\tilde{Z} = \sqrt{W_+} - \sqrt{W_-}. \tag{9.9}$$

With respect to the original version of the RIPPER algorithm, applied criteria, which are based on weights, are the only differences. Thus, again, the training set (equation 9.1) is randomly split into training and test samples. The rule is generated from general to specific with the use of a greedy search, and final sequences of conditions are deleted to minimize equation 9.6. Now weights (equation 9.7) are calculated on the test subsample and the weight of a rule has the a form shown in equation 9.8. Finally, classification is made on the basis of rules that cover the object $\mathbf{x}$:

$$H(\mathbf{x}) = sign\left(\sum_{R_t : \mathbf{x} \in R_t} \tilde{C}_{R_t} \right). \tag{9.10}$$

As in many rule-based models, the order of class descriptions can affect the classification. An unordered set of rules in SLIPPER leads to a lower classification error, but the complexity of the model is much higher (Kubus, 2008). Cohen and Singer (1999) obtained promising results comparing classification errors of SLIPPER with RIPPER and C4.5rules.

9.4 Experiments

Our empirical experiments have been carried out on the benchmark problems from the UCI Machine Learning Repository (Lichman, 2013), which are commonly used for comparative studies. We have chosen data sets that differ in the number of objects, variables, and classes. There are both quantitative and nominal variables, and a few data sets have missing values of variables. When the original test set was not available we randomly split the data into test and training samples in the proportion 1:2. For the calculation we use the implementation of the AdaBoost.M1 algorithm (Freund and Schapire, 1997), which is available in the package adabag for R, and Cohen's original implementation of SLIPPER and RIPPER. We keep the settings in AdaBoost.M1 to the default values suggested in Alfaro et al. (2013). In SLIPPER, the number of iterations is tuned with the use of cross-validation (the numbers from 1 to 100 are checked and the one for which the classification error is the lowest is chosen). Based on the results obtained in Kubus (2008), we use the version of the algorithm that generates an unordered set of rules.

To begin with, we compared the generalization error of the algorithms AdaBoost.M1 and SLIPPER, which apply boosting in different ways. Results are presented in Table 9.1. Adaboost.M1 considerably outperforms SLIPPER in many cases, and SLIPPER yields a lower classification error only five times. We also compared the run times for the three largest data sets (2.1 GHz processor and 4.0 GB of RAM) (see Table 9.2). In fact, the run times are comparable when the number of iterations in SLIPPER is tuned using cross-validation. Setting the number of iterations arbitrarily as 50 we obtained distinctly lower run times for SLIPPER. In such cases classification errors are comparable (Kubus, 2008).

Table 9.1 Classification error (in %) estimated for test samples.

Data sets	SLIPPER	AdaBoost.M1
adult	13.02	15.47
breast cancer	5.26	2.11
car	4.34	9.38
credit australian	12.61	13.91
credit german	23.72	20.72
echocardio	29.55	27.27
ecoli	19.64	13.39
glass	39.06	21.88
heart-disease C	22.77	20.79
hepatitis	13.46	13.46
ionosphere	9.40	7.69
lymphography	10.20	6.12
nursery	5.12	2.89
pima	24.22	24.22
satellite	11.05	11.60
sonar	25.71	18.57
vehicles	26.60	25.89
vowel	17.27	6.36
wine	0.00	1.69
zoo	26.47	8.82

Table 9.2 The comparison of run time (in seconds) for the largest data sets from Table 9.1.

Data sets	Number of objects	Number of variables	SLIPPER run time	AdaBoost.M1 run time
adult	48842	14	24.7	308.6
nursery	12960	8	3.0	69.5
satellite	6435	36	21.4	174.1

The second experiment concerns the problem of stability. Applying boosting for tree-based models guarantees an improvement in stability, so does SLIPPER build more stable models than its predecessor RIPPER? We have chosen three data sets: credit australian, ionosphere and hepatitis. Each of them is randomly split into test and training samples 30 times in the proportion 1:2. The models are built on training sets running both SLIPPER and RIPPER, and classification errors are estimated using test samples. Having 30 errors for each data set and for two algorithms, we check the statistical significance of the differences between variances. SLIPPER never once had a lower variance of the classification error at the significance level of 0.05 (see Table 9.3).

In the next experiment we want to get the answers to two questions. How precisely do the algorithms of interest detect irrelevant variables in the data and how much is the classification error affected by such variables? We used the *Pima* data set, which contains eight quantitative predictors, two classes, and 768 objects (which we randomly split into 512 objects for training and 256 objects for testing). Additionally, we introduce to the original data set 30, 60, or 100 irrelevant variables, which are randomly generated according to the same distribution in both classes. We consider standardized normal distribution, a mixture of normal distributions (1/4 observations from $N(0, 1)$, and the rest from $N(5, 1)$) and the exponential distribution with parameter $l = 2$. Tables 9.4 and 9.5 show the numbers of irrelevant variables introduced to the classifiers and classification errors estimated on the test sets. We have included in this comparison single trees built by the popular CART algorithm (Breiman et al., 1984).

Classification trees introduce no irrelevant variables to the models and they sometimes outperform the ensembles of trees. This may be caused by pruning. We apply

Table 9.3 Standard deviations and test for variances for 30 estimations of the classification error.

Data sets	Standard deviation RIPPER	Standard deviation SLIPPER	F	p
credit australian	1.19	0.42	8.14671	0.06649
ionosphere	0.67	1.25	3.50202	0.25219
hepatitis	1.02	3.44	11.28620	0.03759

Table 9.4 Numbers of irrelevant variables introduced to the classifiers.

	Normal distribution			Mixture			Exponential distribution		
	Number of irrelevant variables			Number of irrelevant variables			Number of irrelevant variables		
Methods	**30**	**60**	**100**	**30**	**60**	**100**	**30**	**60**	**100**
CART	0	0	0	0	0	0	0	0	0
RIPPER	0	1	2	2	0	0	1	4	1
SLIPPER	4	5	4	9	3	2	9	0	9

Table 9.5 Classification error (in %) estimated in test samples for the *Pima* data set with irrelevant variables added from various distributions.

	Normal distribution			Mixture			Exponential distribution		
	Number of irrelevant variables			Number of irrelevant variables			Number of irrelevant variables		
Methods	**30**	**60**	**100**	**30**	**60**	**100**	**30**	**60**	**100**
CART	22.67	21.87	21.87	21.87	21.87	21.87	21.87	21.87	21.87
AdaBoost.M1	22.27	21.87	21.87	21.87	22.66	22.66	21.87	22.27	23.44
RIPPER	20.31	21.09	25.39	25.00	21.09	24.22	23.83	23.83	22.66
SLIPPER	22.27	23.83	23.44	23.83	21.09	23.05	26.95	21.87	28.52

CART with cost-complexity pruning, and we do not prune the trees in the ensemble. For this reason, AdaBoost.M1 introduces a lot of irrelevant variables to the model. We have observed over 2000 nodes with such variables in every ensemble. Note, however, that in many cases the accuracy of the model has not been reduced, or has been reduced slightly. It is surprising that RIPPER, which has implemented advanced pruning techniques, is not able to detect perfectly the irrelevant variables. The worst results, however, have been obtained by SLIPPER. This method is not robust to irrelevant variables, introducing considerably more such variables than RIPPER. This drawback of SLIPPER probably determines the lower classification accuracy in many cases.

9.5 Conclusions

Boosted trees are commonly regarded as one of the most attractive methods of classification due to accuracy. Breiman referred to AdaBoost with trees as the “best off-the-shelf classifier in the world”. Having carried out a few comparisons we would say that adaptation of boosting to the induction rule was not so successful. SLIPPER outperforms its predecessor RIPPER (Cohen and Singer, 1999) and even single classification trees CART (Kubus, 2008) due to accuracy of classification. However, it is not as accurate as AdaBoost.M1, but its advantage is the possibility of interpretation. SLIPPER does not improve stability in comparison to RIPPER. Since it is quite accurate in the same time, SLIPPER could be used as a weak learner. Unfortunately, this boosted rule learner is not robust to irrelevant variables. Thus, the application of SLIPPER should be pre-processed with feature selection.

10

Credit Scoring with Individuals and Ensemble Trees

M. Chrzanowska, E. Alfaro, and D. Witkowska

10.1 Introduction

Decision making, being an integral part of any business, is usually a quite complex process, particularly when the decisions are in the economic sphere and their consequences are considered in terms of profit and loss. Usually, the results of such decisions are not known in advance, which means that they are always accompanied by risk. There are many threats connected with economic activity, and their identification as well as an assessment of their impact on various processes and phenomena has been the aim of numerous studies. In recent years, research on financial issues and methods for identifying, measuring, and minimizing risk in this sector has been developing.

The dynamic growth in the number of loans granted to individual customers requires the right tools that enable processing of applications in a short time and recognizing customers who are not credible, hence the need to seek effective methods that support the credit decisions. Particularly in recent years, following the development of financial markets and data processing systems, more sophisticated methods applying nonlinear models and multidimensional analysis have become more and more popular. In particular, researchers pay special attention to aggregated models, which can be used to evaluate the credit risk (Chrzanowska and Witkowska, 2007a,b; Wang et al., 2012; Louzada et al., 2011; Lahsasna et al., 2010, among others).

The aim of this research is to examine the efficiency of ensemble trees in the classification of a bank's individual clients according to an evaluation of their creditworthiness. In this chapter we compare the results obtained for different assumptions made during the construction of the model. The main differences concern the sample structure, the set of discriminant variables and applied algorithms, and their impact on the quality of dichotomous classification of individual borrowers. The chapter summarizes the authors' previous research on the application of the classification trees to creditworthiness evaluation (Chrzanowska and Witkowska, 2007a,b, 2009; Chrzanowska et al., 2009a,b, among others).

Even though there are several versions of the boosting algorithms, as seen in Chapter 3, we use the most usual AdaBoost algorithm (Freund and Schapire, 1996). These methods have been proved to be very successful, but they have scarcely been used in economic applications up to now (Lemmens and Croux, 2006; Alfaro et al., 2008b).

Ensemble Classification Methods with Applications in R, First Edition.
Edited by Esteban Alfaro, Matías Gámez, and Noelia García.

10.2 Measures of Accuracy

Traditionally, the effectiveness of classification has been measured by the average accuracy, i.e. the proportion of correctly classified examples in the training or test sample. This measure has been shown to be inefficient when the sample is imbalanced (Swets, 1988). For instance, in a problem where there are 95% credible clients and 5% non-credible clients, the default rule, which always assigns the majority class, would have an average accuracy of 95%, which is apparently a good result although it does not predict correctly any of the non-credible clients and is therefore not a desirable classifier. Kubat and Matwin (1997) proposed a special measure to assess the behavior of classifiers in imbalanced samples, the *g-mean accuracy* (*PRG*). To evaluate this measure, it is necessary to define the positive accuracy ratio (*PRCC*) as the percentage of credible clients correctly classified and, in the same way, the negative accuracy ratio (*PRIC*) as the percentage of non-credible clients correctly classified. Then, the g-mean accuracy is the geometric mean of positive and negative accuracies. In the example above, $PRCC = 100\%$ and $PRIC = 0\%$, thus $PRG = 0$, indicating the sloppy performance of the default rule. In our investigation we use four popular accuracy measures:

1. *PRCC* describes the share of properly recognized credible clients:

$$PRCC = \frac{TP}{TP + FN} \tag{10.1}$$

2. *PRIC* describes the share of properly recognized non-credible clients:

$$PRIC = \frac{TN}{TN + FP} \tag{10.2}$$

3. *PRAC* is the average accuracy, i.e. the share of properly recognized average clients:

$$PRAC = \frac{TP + TN}{TP + TN + FP + FN} \tag{10.3}$$

4. *PRG* is the geometric mean of *PRCC* and *PRIC*:

$$PRG = \sqrt{PRCC \cdot PRIC} \tag{10.4}$$

where TP and TN are the numbers of properly recognized clients from the jth class (for $j = 0, 1$), i.e. creditworthy (TP) and non-creditworthy (TN) clients, FP is the number of borrowers who do not pay off credit in time but are recognized by the model as creditworthy clients, FN is the number of borrowers who pay off credits in time but are recognized by the model as non-creditworthy clients.

Since in our analysis we use four different measures of the classification accuracy, which describe the quality of the models in different terms, to make the comparison of synthetic results generated in the experiments provided we also apply the simple average of all measures, i.e.

$$SM = \frac{PRCC + PRIC + PRAC + PRG}{4} \tag{10.5}$$

which lets us compare the performance of the constructed models from all considered points of view.

10.3 Data Description

The research was conducted using actual data regarding the individual borrowers that got a mortgage credit in one of the commercial banks that operate in Poland. The database contains records of 2576 borrowers who signed a contract with the bank between February 1998 and March 2004. Among them 419 clients did not pay off their credit on time. Each borrower is described by 11 features features ($y, x_1 - x_{10}$) that are both qualitative and quantitative:

- y: risk (1 denotes a borrower who repaid the loan on time and 0 denotes a borrower who did not repay the loan on time)
- x_1: client (1 denotes a standard client and 0 a VIP client)
- x_2: gender (1 denotes female and 0 denotes male)
- x_3: currency (1 denotes Polish currency (PLN), y=0 denotes currency other than PLN)
- x_4: age (in years)
- x_5: the place the credit was granted: M1 denotes cities with 500,000 and more inhabitants, M2 denotes cities with 10,000–500,000 inhabitants, and M3 denotes cities with fewer than 10,000 inhabitants)
- x_6: value of the already repaid loan (in PLN)
- x_7: period of repayment (in years)
- x_8: value of credit (in PLN)
- x_9: actual interest rate of credit.

In addition, a new variable has been constructed:

- x_{10}: the change in the credit rate:

$$x_{10} = \frac{x_6}{x_8} \times 100 \tag{10.6}$$

The first stage of research consists of the statistical analysis of all variables and the selection of variables that were significant in the previous investigation (Chrzanowska and Witkowska, 2007a,b). For further analysis the following variables are chosen: x_2 gender, x_4 age, x_5 place of credit granting, x_6 share of the already repaid loan, x_7 period of loan repayment, and x_{10} change in the credit rate.

The next stage consists of analysis of the structure of borrowers and loans. The aim of this stage is to find unusual elements as well as attributes of small variability (quasi-constant). Selected information is presented in Tables 10.1 to 10.5.

Among 2576 borrowers there were 1486 men and 1090 women. It should be noted in Table 10.1 that the majority of borrowers were aged 40–55 and this age group constituted more than 56% of all participants. The smallest percentage of borrowers was found in the older age groups. Borrowers over 65 made up no more than 1% of borrowers. This age group is usually characterized by low income, which, although steady, lowers their creditworthiness. Moreover, elderly people avoid risk and therefore they are not willing to take credit. The youngest borrowers rarely have high income and usually they have no credit history. For this reason, banks reluctantly lend money to this group of clients.

One of the features analyzed for borrowers was the place of credit granting. For this feature three variants, M1, M2, and M3, are distinguished. The feature is characterized by weak concentration. Most credit was granted in small towns and medium-sized cities (Table 10.2).

Table 10.1 Borrowers according to age (x_4).

Age (in years)	Number of borrowers	Frequency (%)	Cumulative frequency (%)
Younger than 30	66	2.53	2.53
30–35	253	9.82	12.35
35–40	315	12.23	24.58
40–45	411	15.95	40.53
45–50	537	20.85	61.38
50–55	515	19.99	81.37
55–65	335	13.00	94.37
60–65	112	4.35	98.72
65–80	33	1.28	100.00
Total	2576	100.00	

Table 10.2 Borrowers according to place credit granted (x_5).

Place of credit granting	Number of borrowers	Frequency (%)	Cumulative frequency (%)
Metropolies (M1)	610	23.68	23.68
Medium cities (M2)	970	37.66	61.34
Small towns (M3)	996	38.66	100.00
Total	2576	100.00	

The majority of borrowers (about 75%) repaid the bank an amount not exceeding 40,000 PLN (Table 10.3). Clients who repaid no more than 10,000 PLN represent 17% of all borrowers. On the other hand, people who repaid a sum between 140,000 and 240,000 PLN constitute only 1.32% of all respondents. It should be underlined that only 14 people (0.54%) repaid an amount higher than 240,000 PLN.

A great variety of customers is observed with regard to the value of the credit. Based on the data presented in Table 10.4, it can be seen that the largest group of borrowers are people who borrowed between 20,000 and 30,000 PLN. Almost half of borrowers did not take credit higher than 40,000 PLN (49.2%). A large number of people (57% of the sample) borrowed a sum between 10,000 and 50,000 PLN. About 15% of borrowers borrowed more than 100,000 PLN.

The last analyzed feature is the agreed time of loan repayment, presented in Table 10.5. This shows that most borrowers (40%) borrowed money for a period of between 14 and 16 years. Another large group is made up of people who intend to repay the loan within 8–10 years. The smallest group is those for whom the deadline for repayment of the loan does not exceed 2 years, as well as those who have been indebted to the bank for longer than 16 years.

The data presented in Tables 10.1 to 10.5 indicate that borrowers who predominate in the database are people aged 40–55 who:

Table 10.3 Borrowers according to share of the loan already repaid (x_6).

Share of the loan already repaid (PLN)	Number of borrowers	Frequency (%)	Cumulative frequency (%)
Less than 10,000	432	16.77	16.77
10,000–20,000	665	25.82	42.59
20,000–30,000	519	20.15	62.74
30,000–40,000	320	12.42	75.16
40,000–50,000	176	6.83	81.99
50,000–60,000	100	3.88	85.87
60,000–70,000	103	4.00	89.87
70,000–80,000	57	2.21	92.08
80,000–100,000	81	3.14	95.22
100,000–120,000	40	1.55	96.77
120,000–140,000	35	1.36	98.13
140,000–160,000	7	0.27	98.40
160,000–200,000	11	0.43	98.83
200,000–240,000	16	0.62	99.45
More than 240,000	14	0.54	100.00
Total	2576	100.00	

Table 10.4 Borrowers according to the value of the credit (x_8).

Value of credit (PLN)	Number of borrowers	Frequency (%)	Cumulative frequency (%)
Less than 10,000	84	3.26	3.26
10,000–20,000	347	13.48	16.74
20,000–30,000	472	18.33	35.07
30,000–40,000	364	14.14	49.21
40,000–50,000	273	10.60	59.81
50,000–60,000	213	8.27	68.08
60,000–70,000	136	5.28	73.36
70,000–80,000	121	4.70	78.06
80,000–90,000	69	2.68	80.74
90,000–100,000	120	4.66	85.40
100,000–110,000	38	1.48	86.88
110,000–120,000	44	1.71	88.59
120,000–130,000	23	0.89	89.48
130,000–140,000	23	0.89	90.37
140,000–150,000	39	1.51	91.88
150,000–200,000	106	4.12	96.00
200,000–250,000	32	1.24	97.24
250,000–500,000	53	2.06	99.30
More than 500,000	19	0.74	100.00
Total	2576	100.00	

Table 10.5 Borrowers according to period of loan repayment (x_7).

Period of loan repayment (years)	Number of borrowers	Frequency (%)	Cumulative frequency (%)
0–2	74	2.87	2.87
2–4	209	7.08	10.67
4–6	376	14.60	25.27
6–8	169	6.56	31.83
8–10	506	19.64	51.47
10–12	113	4.39	55.86
12–14	145	5.63	61.49
14–16	983	38.16	99.65
16–30	9	0.35	100.00
Total	2576	100.00	

- live in small towns and medium cities (M2 and M3)
- received credit of 10,000–50,000 PLN
- repaid the bank an amount not exceeding 40,000 PLN
- agreed to repay the loan within 14–16 years.

10.4 Classification of Borrowers Applying Ensemble Trees

This research consists of many experiments because of:

- the different sets of variables
- the application of boosting and bagging methods
- the diversified structure of the samples concerning both the proportions of credible and non-credible clients together with different proportions of training and testing sets

which are applied in our investigation. It is worth mentioning that selection to the samples was made randomly but separately from two subsets containing credible and non-credible clients (sampling without replacement).

According to our previous investigations, in this study we present two sets containing V1 (four variables, x_5, x_6, x_7 and x_{10}) and V2 (six variables, x_2, x_4, x_5, x_6, x_7 and x_{10}). There are also four variants of samples, denoted S1, S2, S3 and S4, constructed under different assumptions regarding the proportion of creditworthy and non-creditworthy clients (symmetric and asymmetric samples). In addition, three variants of proportions between learning and testing sets (denoted as A, B and C) are distinguished. Taking into consideration how the samples are divided into training and testing sets, we used the testing sets which contain 30% (sample A), 32% (sample B) or 20% (sample C) of the whole sample.

Samples S1 and S2 are symmetric (balanced), which means that they contain the same number of borrowers who repay and do not repay their credits in time. Sample S1 contains 742 objects (in training and testing sets). Assuming that the test sample contains about 30% of clients (variant S1A), the learning set includes 514 observations and the remaining 226 cases form a test set. There are 400 or 500 clients in sample S2, depending on the share of elements used in the training and test samples (B and C, which contain different number of observations). Sample S3 contains 16% of non-credible clients, i.e. as observed in the database obtained from the bank, i.e. there are 2327 clients in variant B and 500 cases in the variant C. Sample S4 contains 9.4% non-credible clients, as observed in Poland in June 2006 (The Polish National Bank Report, 2006). Sample S4B includes 1450 objects, and S4C has 500 clients. In other words, in samples denoted by B, the number of non-creditworthy clients is constant and there are 200 such objects (the same in all samples), i.e. 136 in the training set and 64 in the testing set. In samples denoted C, the number of all elements in the whole sample is constant (each sample contains 500 elements, with 100 elements in the testing sets).

The structure of objects belonging to both classes is maintained for both sets. Thus, the number of elements in the samples changes from 400 to 2327, and the number of non-credible clients differs from 47 to 250 in the single sample while the number of credible clients varies from 250 to 453 (see Table 10.6). It can also be seen that among seven sets of observations only one (S1A) contains the reduced number of variables (i.e., variable set V1).

In our research we applied different methods, providing a series of experiments, but only the "best models", selected according to the generated classification errors, are discussed. Using sample S1A we constructed models applying boosting and bagging algorithms. The former is built using 100 individual trees. The size of each single tree was set using the values $cp = 0.03$ and $maxdepth = 3$ in the respective functions in the R program. The latter, i.e. the aggregated bagging model, was built using 40 single predictors.

Figures 10.1 and 10.2 show the ranking of discriminatory variables in terms of their importance in both models. For the boosting model the most significant predictor is variable x_7 (the period of loan repayment). The importance of other variables (x_5, x_6, x_{10}) is similar, although their discriminatory abilities are smaller. In the bagging model all variables are significant. Nevertheless, variable x_{10} has the highest value in the

Table 10.6 Structure of samples used for further experiments. Note that $y = 1$ denotes the borrower who paid back the loan in time, and $y = 0$ otherwise.

Sample description	S1A		S2B		S3B		S4B	
Variable set	**V1**				**V2**			
	$y = 1$	$y = 0$	$y = 1$	$y = 0$	$y = 1$	$y = 0$	$y = 1$	$y = 0$
Learning set	257	257	136	136	1446	136	850	136
Testing set	113	113	64	64	681	64	400	64
Experiment 2			S2B		S3B		S4B	
			$y = 1$	$y = 0$	$y = 1$	$y = 0$	$y = 1$	$y = 0$
Learning set			200	200	336	64	362	38
Testing set			50	50	64	36	91	9

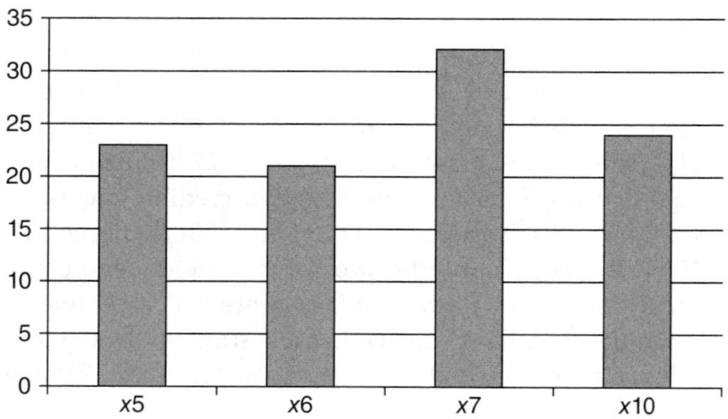

Figure 10.1 Ranking of predictor importance for the boosting model evaluated for sample S1A.

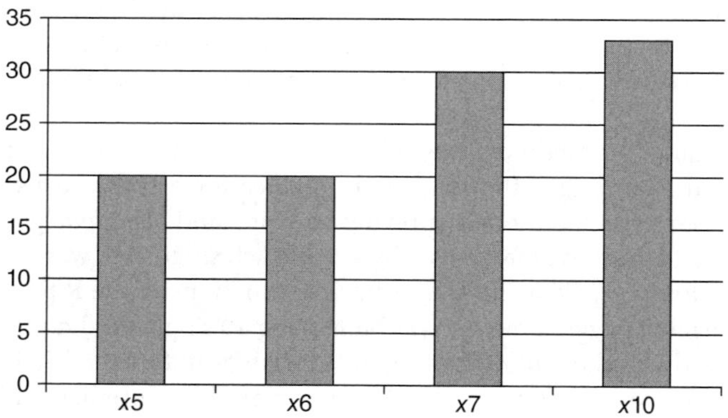

Figure 10.2 Ranking of predictor importance for the bagging model evaluated for sample S1A.

Table 10.7 Results of classification applying boosting and bagging models: the testing set.

		Predicted		
	Observed	$y = 0$	$y = 1$	**Percentage of correctly recognized objects in each class**
Boosting	$y = 0$	92	21	81.42
	$y = 1$	29	84	74.34
Bagging	$y = 0$	7	26	76.99
	$y = 1$	27	86	76.11

ranking (the highest potential for discrimination). However, variable x_7 also obtained a high value.

Table 10.7 contains the results of the borrower classification obtained for sample S1A applying both models considered. Using the boosting model, 78% of objects were correctly identified in the testing set ($PRAC = 0.78$). This model also correctly recognized 81% of lost loans and 74% of regularly paid loans in the learning sample. The bagging model properly recognized 77% of non-creditworthy clients and 76% of creditworthy

Table 10.8 Comparison of accuracy measures for the training and testing sets.

	Boosting model	Bagging model
	Training set	
PRCC	80.54	65.37
PRIC	84.82	71.60
PRAC	82.68	77.88
PRG	82.69	68.41
	Testing set	
PRCC	74.34	76.11
PRIC	81.42	76.99
PRAC	77.88	76.55
PRG	76.50	76.55

borrowers. Analyzing the results, it can be seen that the results of classification using both models are very similar. The boosting model has a slight advantage over the bagging one in general classification but the latter seems more efficient in identifying unreliable borrowers.

Analyzing the results of borrower classification (Table 10.8), we notice that the efficiency in recognition of the bank clients provided by both models is similar, and there is not a big difference between the training and testing sets.

Further research is provided for the set V2 containing six variables and the other samples described previously. The experiments are carried out using ensemble classification trees built with the adabag library (Alfaro-Cortes et al., 2018; Alfaro et al., 2008b) for the R Cran program. The boosting models are constructed on the basis of 200 single classifiers, while the bagging models are constructed using 40 single trees, with values selected on the basis of several trials. (The size of each single tree was set using the values $cp = 0.03$ and $maxdepth = 3$ in the respective functions in the R program.)

As can be seen in Tables 10.9 and 10.10, the boosting algorithm presents a similar performance both in the training and testing sets. The average accuracy seems an illusory improvement of the prediction ability in samples S3B and S4B, owing most to the *PRCC*, which achieves nearly 100% of correctly classified examples. On the other hand, the *PRG* gives a more realistic view because taking into account the decline of *PRIC* confirms the worst classification results for imbalanced classes.

If the accuracy of the model is described by the results obtained for the testing sets (Tables 10.8 and 10.10), it is clear that credible clients are better recognized by the algorithms using asymmetric samples (S3 and S4) since these samples contain more observations concerning creditworthy clients, whereas non-credible clients are usually better recognized when balanced samples are used. The biggest classification errors are observed in testing sets for experiments S3C and S4B for both boosting and bagging models together with S3B for bagging.

Table 10.9 Comparison of accuracy measures for the training samples.

Boosting model						
Samples	S2B	S2C	S3B	S3C	S4B	S4C
PRCC	91.18	78.50	96.12	99.40	99.79	100.00
PRIC	83.09	67.50	32.35	51.56	52.21	84.21
PRAC	87.13	73.00	87.32	91.75	95.70	98.50
PRG	87.04	72.79	55.76	71.59	72.18	91.77
Bagging model						
Samples	S2B	S2C	S3B	S3C	S4B	S4C
PRCC	77.94	82.50	99.18	98.21	99.38	98.90
PRIC	61.76	64.50	32.35	40.63	30.15	44.74
PRAC	69.85	73.50	89.96	89.00	93.43	93.75
PRG	69.38	72.95	56.64	63.17	54.74	66.52

Table 10.10 Comparison of accuracy measures for the testing samples.

Boosting model						
Samples	S2B	S2C	S3B	S3C	S4B	S4C
PRCC	89.06	94.00	99.50	100.00	99.27	98.90
PRIC	81.25	58.00	45.31	33.33	39.06	77.78
PRAC	85.16	76.00	92.03	76.00	94.09	97.00
PRG	85.07	73.84	67.15	57.74	62.27	87.71
Bagging model						
Samples	S2B	S2C	S3B	S3C	S4B	S4C
PRCC	70.31	94.00	98.75	98.44	98.53	98.90
PRIC	56.25	64.00	23.44	33.33	25.00	66.67
PRAC	63.28	79.00	88.36	75.00	92.21	96.00
PRG	62.89	77.56	48.11	57.28	49.63	81.20

Taking into consideration the proportion of the number of objects included in the testing and training sets (B and C), it is hard to find any relation between these cases. However, for bagging models better results are obtained for variant C in terms of *PRG*, i.e. when the proportion of training and testing sets is 80/20. Applying the boosting algorithm, such a relation is observed only for sample S4.

Comparing the results obtained for both variable sets using balanced samples (S1A and S2B) it is clear that the boosting algorithm performs better when the variable set V2 is used and the bagging model performs better for the set V1. We also notice that the proportion of the sample counts influences the classification accuracy since when comparing results obtained for samples S2B and S2C, which contain significantly different

Table 10.11 Comparison of synthetic measures.

		Training		Testing	
Samples	**Variable set**	**Boosting**	**Bagging**	**Boosting**	**Bagging**
S1A	V1	82.68	70.82	77.86	76.55
S2B	V2	87.11	69.73	85.14	63.18
S2C		72.95	73.36	75.46	78.64
S3B		67.89	69.53	76.00	64.67
S3C		78.58	72.75	66.77	66.01
S4B		79.97	69.43	73.67	66.34
S4C		93.62	5.98	90.35	85.69

numbers of observations, to samples S1A and S2C, which contain similar numbers of observations, our conclusion is the opposite.

In terms of the synthetic measure *SM* (equation 10.5) boosting models generate better results than bagging models, except for sample S2C in the training and testing procedures, whereas in the training process the bagging model generates better results for sample S3B (see Table 10.11). Generally speaking, all results seem to be very similar. However, in terms of synthetic measure evaluated for the testing procedure, the best results are obtained for sample S4C and the worst are generated by the boosting model for S3C and the bagging algorithm for S2B.

10.5 Conclusions

The aim of the study was to find out if application ensemble trees let us properly recognize credible and non-credible individual bank clients and assess the impact of the sample structure on the quality of dichotomous classification of individual borrowers. What seems particularly interesting is the answer to the question of whether recognition of unreliable customers, whose identification is of particular interest to banks (and a sample usually includes fewer unreliable customers than reliable ones), is better for a sample that includes the same number of objects from both classes.

Recognition of individual clients in creditworthiness evaluation is a difficult task since the problem of loan repayment concerns the future while the classification of clients is made on the basis of information regarding past and present situations, therefore mistakes always appear. However, a question arises as to how to distinguish between large and small errors. Let us assume that the accepted error is equal to that made by the bank which properly recognized 83.73% clients, which is the share of loans repaid in time. It is worth mentioning that in the creditworthiness procedure the bank used more information than is used in the model construction and the borrower classification.

Making such assumption we notice that, for 30 analyzed models, *PRAC* is bigger than in the testing procedure only for samples S4B and S4C for both boosting and bagging models, and additionally for the boosting algorithm for samples S2B and S3B. However, if *PRCC* is taken into account, all samples but S1A in boosting models and all samples but S1A and S2B in bagging models generate better results in the recognition of clients than the procedure provided by the bank.

In conclusion, we can say that all presented models are efficient in the classification of clients, therefore ensemble trees may help financial institutions with the creditworthiness evaluation of individual clients.

The boosting algorithm outperforms the bagging algorithm in terms of the average accuracy in all test samples except samples S1 and S2C. In addition, boosting also beats bagging models regarding the g-mean accuracy except for test sample S2C. It has been shown that dealing with imbalanced classes, the geometric mean accuracy gives a more realistic measure than the traditional and better-known average accuracy. These results are in line with previous studies related to imbalanced classes.

11

An Overview of Multiple Classifier Systems Based on Generalized Additive Models

K.W. De Bock, K. Coussement, and D. Cielen

11.1 Introduction

Humans tend to seek the advice of other people when faced with an important decision and combine individual opinions into a final decision. While this process is natural to humans, the advantages of such an automated approach have only recently been discovered by the computational intelligence community (Polikar, 2006).

This approach, known as multiple classifier systems, combines the decisions from individual classifiers to obtain a more accurate classifier. Experimental research confirms significant accuracy improvements from such an approach over using the prediction from a single classifier (Bauer and Kohavi, 1999; Breiman, 1996b; Kohavi and Kunz, 1997). Multiple classifier systems are also known as ensemble methods, committee of classifiers, and mixture of experts (Polikar, 2006).

Three popular ways of creating the individual classifiers for multiple classifiers systems are bagging, random subspace modeling, and boosting. In bagging or bootstrap aggregation (Breiman, 1996a), multiple models trained on a different subspace of observations are aggregated. Random subspace modeling (EAM; Ho (1998)), also known as feature bagging, uses a random sample of features instead of observations. Similar to bagging, boosting (Schapire, 1990) also resamples the observations but in contrast to bagging observations get a strategically chosen weight. Examples of well-known ensemble algorithms are random forests (Breiman, 2001) and ADABoost (Freund, 1995; Freund and Schapire, 1997). Random forests, a special case of bagging, uses decision trees as a base classifier. ADABoost or adaptive boosting is a popular boosting algorithm that gives more weight to observations that are misclassified in previous iterations.

After training the individual classifiers, they must be combined into a new classifier. The two main categories of methods for combining the individual classifiers are weighting and meta-learning methods.

A central choice the researcher has to make when using multiple classifier systems is the choice of what base classifier to use. Many algorithms have been studied in this regard, such as decision trees (Breiman, 2001), k-nearest neighbors (Bay, 1999), support vector machines (Wang et al., 2009), and many others. This chapter reviews the GAMens family of multiple classifier systems (De Bock et al., 2010) that use general additive models (GAMs; (Hastie and Tibshirani, 1986)) as a base classifier.

Ensemble Classification Methods with Applications in R, First Edition.
Edited by Esteban Alfaro, Matías Gámez, and Noelia García.

We will show that GAM-based multiple classifiers serve as a good extension to the ensemble-based literature because they are stronger predictors than a single GAM-based classifier, are as strong as other multiple classifier systems, allow for interpretability, and have proven to be effective in many real-world applications.

To expound this statement, this chapter starts by introducing the theory behind GAMens classifiers, and then looks at real-world applications and how GAM-based multiple classifier systems compare to other popular algorithms. We then introduce the GAMens package for R (De Bock et al., 2018), which allows the reader to experiment with this novel technique.

11.2 Multiple Classifier Systems Based on GAMs

Before studying the GAMens classifier, we first refresh the concepts behind GAMs as they are the cornerstone of the GAMens multiple classifier system then introduce three types of GAMens classifier. Finally, the GAMensPlus classifier is introduced, which gives enhanced interpretability to researchers. After reading this chapter the reader should have a solid understanding of the theory and concepts behind the GAMens family of classifiers and be equipped to start using classifiers from the GAMens family.

11.2.1 Generalized Additive Models

The parametrization of GAMens classifiers builds partially on the characteristics of ensemble methods and partially on the characteristics of a GAM classifier, therefore the reader should become acquainted with the theory of GAMs before going ahead to work with GAMens classifiers.

GAMs are a generalization of generalized linear models (GLMs) proposed by Hastie and Tibshirani (1986). GAMs use a sum of smooth instead of a linear functional form and have successfully proven their ability to capture nonlinear relationships between explanatory variables and a response variable in many domains (Berg, 2007; Clements et al., 2005; Kawakita et al., 2005).

GAMs extend the framework of GLMs (Mccullagh and Nelder, 1972), which makes up a broad range of parametric regression models characterized by

(i) a response variable belonging to any distribution within the exponential family (the random component)
(ii) a fixed function that represents any functional relationship between the combined linear effect of the predictors and the expected value of the outcome (the link function)
(iii) the assumption of a combined linear effect of the explanatory features (the systematic component) (Lado et al., 2006).

In GAMs proposed by Hastie and Tibshirani (1986), the influence of an explanatory feature is no longer subject to any linear or other parametric specification, but instead fit using an arbitrary non-parametric function. GAMs replace the linear combination $\sum_{k=1}^{p} \beta_k \mathbf{X}_k$ with the additive form $\sum_{k=1}^{p} f_k(\mathbf{X}_k)$, where each partial function f is an unspecified smooth function. To accommodate a binary response variable and the inclusion of categorical variables, the GAM specification that is considered in this study is a logistic, semi-parametric additive model:

$$\text{logit}(P(Y=1|X)) \equiv \log\left(\frac{P(Y=1|X)}{1-P(Y=1|X)}\right) = \sum_{j=1}^{p_c} s_j X_j + \sum_{k=1}^{p_b} \beta_k X_k \tag{11.1}$$

where features X_j, $j = 1, \ldots, p_c$ are continuous variables, X_k, $k = 1, \ldots p_c$ are dummy-coded components of categorical variables and the smooth functions $s_1(X_1), s_2(X_2), \ldots, s_{p_c}(X_{p_c})$ are smoothing splines that estimate the non-parametric trend for the dependence of the logit on $X_1, X_2, \ldots X_{p_c}$.

A smoothing spline for variable X solves the following optimization problem: amongst all functions $\eta(x)$ with continuous second-order derivatives, find the function that minimizes the penalized residual sum of squares via

$$\sum_{j=1}^{n} y_i - \eta(x_i)^2 + \lambda \int_a^b (\eta''(t))^2 dt \tag{11.2}$$

where λ is a fixed constant and $a \leq x_1 \leq x_2, \ldots, \leq, x_n \leq b$. The goodness of fit is measured by the first part of equation 11.2, while the second term is a penalty term that penalizes curvature in the function, where the degree of penalization is determined by the smoothing parameter λ. The complexity of $\eta(x)$ is measured by λ, which is inversely related to the degrees of freedom (df). If it is small (i.e., the df are large), $\eta(x)$ is any function that approaches an interpolation to the data. When λ is large (i.e., the df are small), $\eta(x)$ is closely related to a simple least squares fit. It can be shown that an explicit and unique minimizer for equation 11.2 exists, i.e. a natural cubic spline with knots at the unique values of x_i (Hastie and Tibshirani, 1990).

11.2.2 GAM-Based Multiple Classifier Systems

De Bock et al. (2010) showed that the GAM classifier can have improved accuracy when are used as a base classifier for a multiple classifier system. They developed ensemble methods using GAMs that can be based on bagging, random subspace modeling or both, and named the classifiers GAMbag, GAMrsm, and GAMens, respectively.

The GAM ensemble algorithms, for which the pseudo code is shown in code listing 1, require the specification of several input parameters. A first set of parameters specifies the ensemble strategy. GAMbag incorporates bagging, which requires parameter b to be true. GAMrsm only implements the random subspace method and requires parameter s to be true. For GAMens, both parameters are set to true. Secondly, the m parameter designates the number of desired GAM base classifiers to be included in the ensemble classifier. Thirdly, the desired number of variables to be selected as random feature subspaces is required (r parameter). Finally, specification of the number of degrees of freedom to be used in the smoothing spline estimation is required (df parameter).

In the prediction phase, outputs of the ensemble member GAMs are combined into an ensemble prediction $C(x)$ by means of average aggregation (or mean combination rule), which is used in many well-known ensemble classifiers (e.g., RSM and rotation forest). In GAM-based ensembles, an ensemble prediction for a given observation takes the average of the posterior class membership probabilities produced by the individual ensemble members.

Listing 11.1: GAMbag. GAMrsm, and GAMens algorithms
Input:

- D: a training set, $D = \{(\mathbf{x}_i, y_i)\}_{i=1}^{n}$; $\mathbf{x}_i \in \mathbb{R}^p$; $y_i \in Y = \{0; 1\}$
- m: number of GAMs in the ensemble
- r: number of randomly selected variables, $r \leq p$
- df: number of degrees of freedom used to estimate smoothing splines $s_1(\cdot), s_2(\cdot), \dots, s_p(\cdot)$
- b: true for bagging (GAMens and GAMbag)
- s: true for RSM (GAMens and GAMrsm)

Training phase
For $i = 1, 2, \dots, m$

- IF S then $R_i = D$ with subset of r randomly selected from $\mathbf{X}$.
- Else $R_i = D$.
- If b D_i, bootstrap sample of R_i.
- Else $D_i = R_i$.
- In D_i, identify continuous variable set $R_{i,c}$ and binary variable set $R_{i,b}$. The numbers of elements of both sets are indicated by $p_{i,c}$ and $p_{i,b}$.
- Estimate lth base classifier as C_l as a semi-parametric GAM with logistic link function and df degrees of freedom for smoothing splines $s_1(\cdot), s_2(\cdot), \dots, s_p(\cdot)$

$$C_l \,:\, P_l(Y = 1|\mathbf{X}) = 1 \bigg/ \left\{ 1 + \exp\left(-\left(\sum_{j=1}^{p_{l,c}} s_{l,j}(\mathbf{X}_j) + \sum_{k=1}^{p_{l,b}} \beta_{l,k}\mathbf{X}_k \right)\right)\right\}$$

with $\mathbf{X}_j \in \mathbf{R}_{l,c}$ and $\mathbf{X}_k \in \mathbf{R}_{l,b}$.

Prediction phase
The probability for observation x to belong to class 1, predicted by ensemble classifier C, is

$$C(x) = \frac{1}{m} \sum_{l=1}^{m} C_l(x)$$

11.2.3 GAMensPlus: Extending GAMens for Advanced Interpretability

In practice, many situations require the ability to interpret the importance of a feature and have knowledge about the confidence in a prediction in a certain input interval. As this is not addressed in the GAMens, GAMbag and GAMrsm methods, De Bock and Van den Poel (2011)) enhanced the GAMens classifier with the model interpretability and named the new algorithm GAMensPlus.

The GAMensPlus algorithm combines the training and prediction phases of GAMens with an explanation phase in which two heuristics are introduced to allow model interpretation: generalized feature importance scores and bootstrap confidence intervals for smoothing splines. This allows decision makers to better understand the influence and relative importance of descriptive features.

Generalized feature importance scores are based on permutation accuracy importance which is the most advanced and reliable variable importance measure available in random forests (Breiman, 2001; Strobl et al., 2007). The generalized feature importance scores in GAMensPlus are in fact generalizations of the permutation accuracy importance used in random forests as they allow a general performance metric to be used instead of only accuracy. As decision makers do not always care for pure accuracy, this broadens the usability of the feature importance scores in real-world applications.

Generalized feature importance scores also do not inherit two deficiencies from permutation accuracy importance (De Bock and Van den Poel, 2011). As the generalized feature importance scores in GAMensPlus use GAMs as base classifiers they do not inherit biased feature selection from CART decision trees, which are the base classifier for random forests. The GAM specification for GAMensPlus also only allows continuous and binary features, and thus avoid the bias introduced by including more than two categories, as shown by Strobl et al. (2007).

Next to feature importance, GAMensPlus introduces (bootstrap) confidence intervals for smoothing splines and regression coefficients for the dummy features. These allow model users to identify the relationship between the predictive feature and the dependent variable plus they show the precision of the identified relationship in particular regions within the range of feature values.

GAMensPlus uses bootstrap percentile intervals (Efron, 1982). To identify the 95% bootstrap confidence interval of the smoothing splines for feature $\mathbf{X}_d$ at each value $x_{d,i}$ of $\mathbf{X}_d$, the empirical distribution of $s_d^*(x_{d,i})$, the random variable of bootstrapped smoothing splines must be identified. A bootstrap confidence interval is then constructed by points at the 2.5th and 97.5th percentiles of this empirical distribution. Repeating this process for each unique value of $\mathbf{X}_d$ then results in the 95% confidence band of the non-parametric regression line (Efron and Tibshirani, 1994). A similar approach is applied to obtain bootstrap confidence intervals for regression coefficients of the dummy features.

11.3 Experiments and Applications

To confirm the usability of the GAMens-based methods, we need to experimentally validate them in a number of settings and see if they indeed have comparable results to similar multiple classifier systems. This part of the text reviews the literature for applications to support our claim that GAMens classifiers are (or are not) valid alternatives for well-known algorithms such as random forests.

11.3.1 A Multi-Domain Benchmark Study of GAM-Based Ensemble Classifiers

A first validation of GAMens comes from placing GAM-based ensemble classifiers next to already existing studies by using the same data sets and algorithms to compare to as earlier studies. To enable comparison, a fixed battery of competitive algorithms, data sets, and statistical tests are used. This section discusses how GAM-based ensemble classifiers perform in these standardized tests.

To this end, De Bock et al. (2010) studied the performance of the GAMens, GAMbag, and GAMrsm classifiers against each other, the pure GAM classifier, and four other ensemble methods: bagging, RSM, random forests tree and rotation forests. These are well-known algorithms often used in classifier benchmark studies (Bauer and Kohavi, 1999; Rodriguez et al., 2006) which allow us to understand the capabilities of GAMens better.

To enhance comparability of the performance of GAMens-based classifiers, the authors use 12 two-class classification data sets from the UCI repository to compare the performance of the different classifiers. This is in correspondence with studies that do similar classifier performance testing (Rodriguez et al., 2006; Zhang and Zhang, 2008). The data sets are also publicly available so that researchers can easily duplicate their findings and compare their own algorithms against them.

To compare the GAMens-based classifiers against other classifiers, the authors used a method based on best practices found in the literature. The performance between the algorithms is benchmarked based on a 5×2-folder cross-validation and the performance metric chosen is area under the receiver operating curve (AUC) as several authors argue that this is an objective performance criterion (Langley, 2000; Provost et al., 1998). To show a significant difference between the classifiers, the non-parametric Friedman test (Friedman, 1937, 1940) is used as suggested by Demsar (2006).

The results of the non-parametric Friedman test are shown in Table 11.1. From this we can conclude that GAMbag performs better than (single) GAM-based classifiers and equally as well as the other ensemble methods. While GAMrsm and GAMens

Table 11.1 Average rank difference (CC-BA) between GAM ensemble and benchmark algorithms based upon De Bock et al. (2010) ($^*p < 0.10$, $^{**}p < 0.05$).

	GAM	RSM	Bagging	Random forest	Rotation forest
GAMbag	−1.92*	−0.25	−0.58	1.33	0.92
GAMrsm	−3.50**	−2.25**	−2.50**	−0.75	−1.00
GAMens	−3.33**	−2.08**	−2.25**	−0.50	−0.83

outperform single GAM, RSM and bagging and do equally well as random forests and rotation forests on the 12 data sets.

Based on this work, we can conclude that the GAM-based multiple classifier systems are competitive with other well-known algorithms such as random forests. In the next section we study how GAMens classifiers perform in a real-world churn case.

11.3.2 Benchmarking GAM-Based Ensemble Classifiers in Predictive Customer Analytics

Algorithms do not live in a laboratory setting and thus must perform well in real-world case studies before they can claim to be a valid alternative to well-proven algorithms. By studying the performance of GAMens classifiers in real-world applications, we gain a better understanding of the usability of GAMens-based algorithms and add to the evidence that they are a valid contender for other ensemble-based classifiers.

Customer churn prediction is such a real-world case that is of importance to many companies. As companies invest a lot of effort and money in acquiring their customer base, they want to know whether a customer is going to end their customer relationship as soon as possible. This enables them to take action to retain the customer, such offering them discounts or other benefits proactively. To predict when a customer is at risk of leaving the customer base, churn prediction is used. GAM-based ensemble classifiers have been studied in the context of churn prediction (Coussement and De Bock, 2013; De Bock and Van den Poel, 2011) and we use their findings here to show the usefulness of GAMens classifiers in this setting.

In one study, Coussement and De Bock (2013) studied the performance of GAM-based classifiers in the context of churn predictions in the online betting industry. They used a data set on actual gambling behavior delivered by BWIN Interactive Entertainment, an online gambling operator. This data set contains the recorded internet bettings of a cohort of 3729 gamblers who played at least four times on three different dates and opened their first account on BWIN during February 2015. A customer is flagged as a churner when he has had no activity for four consecutive months. The churn-incidence is 30.64% for the in-sample data set (2486 players) and 23.74% for the out-of-period data set (1243 players).

The authors calculated the top-decile lift (TDL) and the lift index (LI) to evaluate the quality of the churn predictions. 5×2 cross-validation is used to calculate the evaluation metrics and results are reported for the in-sample test set (TEST) and the out-of-sample test set (OOP). In total 10 runs were done. The results of each run, together with the average performance over 10 runs and the standard error, are considered for both TDL and LI.

Table 11.2 summarizes the average performance of the different classifiers and checks statistical significance based on the non-parametric Wilcoxon-signed rank test, which ranks the differences in performance over the 10 cross-validation runs, ignoring the signs, and comparing the signs for positive and negative differences (Demsar, 2006; Wilcoxon, 1946). The Bonferroni correction is used to correct the problem of multiple corrections and to consider the family-wise error rate.

From Table 11.2, they concluded that ensemble methods outperform single methods in the online gambling setting and that GAMens and random forests do not statistically differ in significance. They also noted that GAM and CART are more sensitive to input

Table 11.2 Summary of the average performance measures over the 5 × 2-fold cross-validation based on Coussement and De Bock (2013). Note: In any rows, performance measures that share a common subscript are not significantly different at $p < 0.05/6$.

		Algorithm			
Evaluation metric	**Set**	**CART**	**RF**	**GAM**	**GAMens**
TDL	TEST	2.2454_a	2.8245_b	2.4586_a	2.7820_b
TDL	OOP	2.9631_a	3.4694_b	2.7524_a	3.4014_b
	TEST	0.7109_a	0.7574_b	0.7178_a	0.7530
	OOP	0.7260_a	0.7775_b	0.7071_a	0.7772_b

changes than their ensemble counterparts GAMens and random forests as the standard errors for CART and GAM are systematically higher than those of random forests and GAMens.

In a second paper, De Bock and Van den Poel (2011) studied the performance of the GAMensPlus classifier in the context of churn predictions. They compare GAMensPlus to a selection of benchmark algorithms on data sets for six real-life customer churn prediction projects on four metrics: accuracy, AUC, top-decile lift and lift index. The benchmark algorithms are bagging, random forests, RSM, logistic regression and GAM.

Experimental results are all based on 5 × 2 cross-validation and the significance of the difference among classifiers is tested using a non-parametric Friedman test followed by Holm's procedure (Holm, 1979) to make post-hoc pairwise comparisons between GAMensPlus and the benchmark algorithms.

Table 11.3 reports the resulting average algorithm rankings and post-hoc Holms test, which confirm the highly competitive performance of GAMensPlus. In terms of accuracy, the strongest competitor is the random forest but GAMensPlus outperforms random forests in terms of AUC, top-decile lift and lift index at the 90% significance level. GAMensPlus systematically outperforms GAM, RSM and bagging.

The previous studies add evidence to the statement that the GAMens family is a valid alternative to other algorithms. The papers reviewed here studied the behavior

Table 11.3 Average algorithm rankings and post-hoc test results (Holm's procedure) based on De Bock and Van den Poel (2011) (*$p < 0.10$, **$p < 0.05$).

	Average rank test			
Algorithm	**Accuracy**	**AUC**	**Top-decide lift**	**Lift index**
GAMensPlus	2.33	1.50	1.58	1.50
Bagging	3.83	4.00**	4.33**	4.50**
Random forest	4.33**	3.5*	3.33*	3.67*
RSM	1.83	433**	4.67**	4.17**
Logistic regression	4.00*	3.33	2.92	3.0*
GAM	4.67**	4.33**	4.17**	4.17*

of the GAMens family in real-world customer churn projects and found that they are competitive with well-known algorithms such as random forests and consistently dominate single classifiers. On top of that, they confirmed that ensemble classifiers are more robust predicters than their single classifier counterpart as they are less sensitive to input changes. This can be of real importance to managers as they need to be able to make decisions and make investments based on the output of an algorithm. They do not always have the option of changing decisions or changing their position frequently.

11.3.3 A Case Study of GAMensPlus used for Customer Churn Prediction in Financial Services

In a real-world setting, managers may need to understand the reliability of a prediction and the importance of the different predictors. We investigate here how GAMensPlus gives researchers and managers this information and thus increases the usability of this algorithm in these settings.

De Bock and Van den Poel (2011) used the bank data set featured in the aforementioned churn prediction study to show how the results of GAMensPlus can be interpreted. For this they calculated the feature importance scores as well the AUC as the top decile lift, as presented in Table 11.4.

Table 11.4 The 10 most important features with feature importance scores based on AUC and TDL based on De Bock and Van den Poel (2011).

	AUC		Top decile lift	
Rank	Feature	Feature importance score	Feature	Feature importance score
1	Checking account balance	0.1276	Recency based on change in balance amount	0.3488
2	Average amount of credit transactions	0.1177	Number of overdraft days (balance < 0 €)	0.3258
3	Number of debit transactions	0.1106	Number of days that overdraft interest is due	0.2754
4	Recency based on last transaction	0.0963	Recency based on change in balance amount	0.2613
5	Recency based on change in balance amount	0.0910	Number of bank accounts	0.2530
6	Total accounts balance	0.0853	Number of checking accounts	0.2241
7	Total number of credit transactions	0.072 1	Average overdraft amount	0.2188
8	Number of bank accounts	0.0674	Average amount of debit transactions	0.2146
9	Number of checking accounts	0.0611	Number of debit transactions	0.2086
10	Total credit movement	0.0599	Checking account balance	0.1815

The list of indicators contains mainly RFM variables (Cullinan, 1977), which summarize a customer's purchase history through quantifying recency (R), frequency (F) and monetary value (M). These features have been identified as strong predictors in customer churn prediction modelling (Kim and Yoon, 2004; Lemmens and Croux, 2006). Many of the features selected based on AUC or TDL are the same but are ordered differently, reflecting the different performance metric underlying the feature importance calculation.

Figure 11.1 shows the 95% bootstrap confidence intervals and average trends for a selection of predictive features. It also includes a histogram that provides an indication

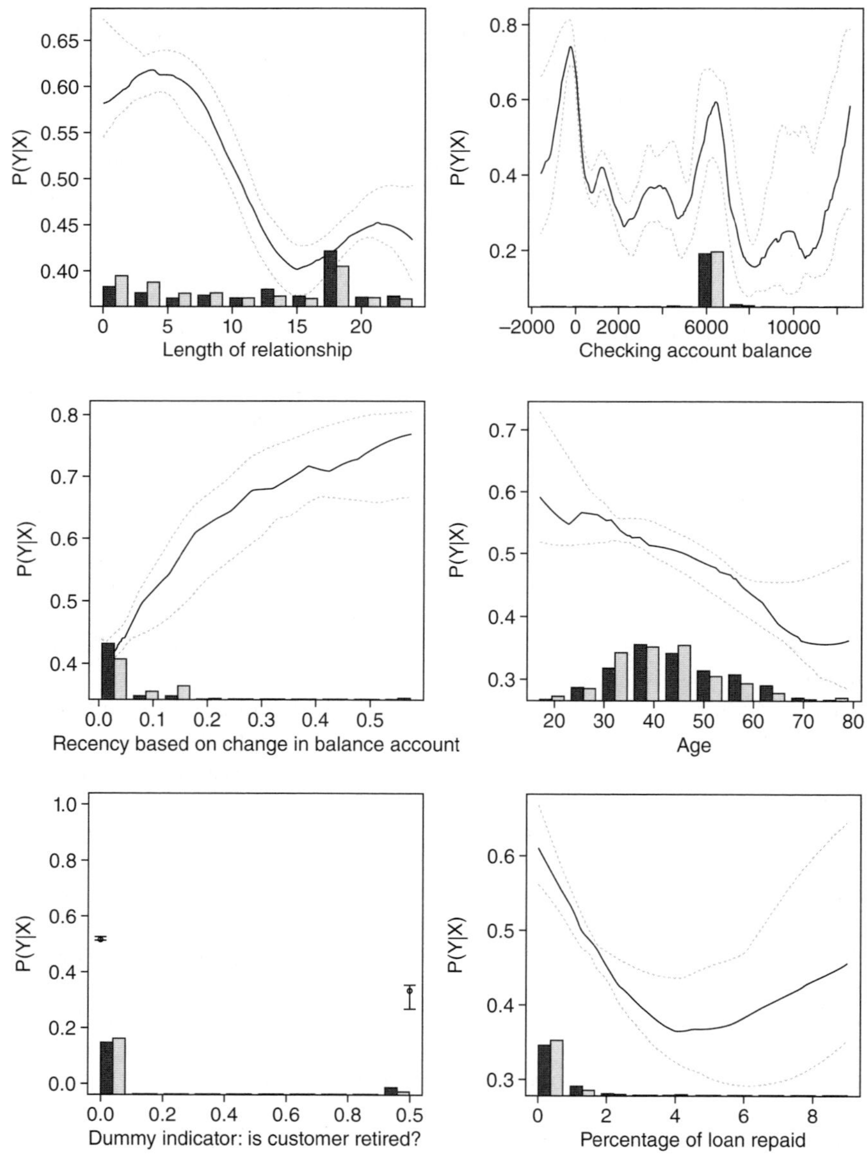

Figure 11.1 Bootstrap confidence intervals and average trends for a selection of predictive features (from De Bock and Van den Poel (2011)).

of the data density at a particular region of a feature. Dark-colored bars represent frequencies for non-churners, while light-colored bars represent frequencies for churning customers. The average trend represents the overall relationship of a feature to the probability to churn, while the confidence bands indicate the reliability of the trend at different regions of the feature range.

Consider, for instance, the age feature, which shows an almost linear and negative relationship between age and the chance to churn, in other words the older you get the less likely you are to churn. On the other hand, there is a nonlinear but positive relationship between a recent change in a bank account and the chance of churning, indicating that the longer you do not use an account the more likely it is that you will churn. The relationship between the amount in the bank account and the chance to churn is nonlinear. In general, the confidence band gets wider in regions with fewer data points.

This study shows the ability of the GAMens classifier to extract important information such as feature importance and confidence interval in the input range of an independent variable. Furthermore, it shows the importance of having the ability to choose a relevant accuracy measure as it influences the ranking of the feature importance.

11.4 Software Implementation in R: the GAMens Package

Gaining hands-on experience with GAMens is possible through the GAMens package (De Bock et al., 2018), an R-implementation of the GAMens-based classifier available under an GPL-2 License.

This package contains three methods that enable researchers and developers to apply the GAMens methods in their studies and projects: GAMens, GAMens.cv, and predict. GAMens. The GAMens method fits GAMbag, GAMrsm or GAMens ensemble methods for binary classifications, GAMens.cv adds cross-validation to the GAMens method, and the predict.GAMens function allows the user to score new data against a trained GAMens model. Predict allows you to score new data against a trained GAMens.

11.5 Conclusions

In this chapter we reviewed the literature around GAMens. First we reviewed the theoretical concepts underpinning the different GAM-based multiple classifier systems: GAMens, GAMbag, and GAMrsm. We also discussed GAMensPlus, an extension of GAMens that adds interpretability of the model to the performance of the other GAMens methods. We then revised the performance of the different GAM-based classifiers against other popular classifiers and their application in real-world cases. Finally, the GAMens package in R was described.

Based on this systematic review, we conclude that GAM-based multiple classifier systems serve as a good extension to the ensemble-based literature because they are stronger predictors than a single GAM-based classifier, are as strong as other multiple classifier systems, allow for interpretability, and have proven to be effective in many real-world applications.

The GAM-based multiple classifier systems did indeed consistently and significantly outperform a pure GAM classifier in all the reviewed papers. In addition to having a better predictive accuracy, the predications are more stable than a pure GAM classifier as the predictions are not so sensitive to small input variations. GAMensPlus allows researchers to study the effect of predictors and gives them a confidence interval. On top of that, GAMens-based classifiers allow the use of a custom performance metric, which increases the applicability of GAMens in to the practical field.

While the GAMens-based classifier did indeed perform well in the real-world cases and proved to be a valid contender for other ensemble-based methods, more studies are needed to confirm this finding as the study towards GAMens-based classifiers is young and more real-world cases need to confirm the findings in the reviewed studies. This would allow our conclusions to be confirmed and transferred to new domains.

Furthermore, the theory around GAMens-based classifiers has many possibilities for expansion. Future research areas could investigate, for instance, more features of ensemble methods, such as new aggregation metrics and ways of generating a battery of single classifiers to feed a multiple classifier systems. To encourage this, researchers can rely on the GAMens package in R to facilitate their research. Nonetheless, the GAM-based multiple classifier system literature and its applications would benefit from having more implementations available to them. This would encourage researchers to test GAMens in multiple real-world settings and include them in comparative studies.

References

Agarwal, S., Graepel, T., Herbrich, R., Har-Peled, S., and Roth, D. (2005). Generalization bounds for the area under the ROC curve. *Journal of Machine Learning Research*, 6: 393–425.

Ahachad, A., Álvarez Pérez, L., and Figueiras-Vidal, A. R. (2017). Boosting ensembles with controlled emphasis intensity. *Pattern Recognition Letters*, 88(Supplement C): 1–5.

Alfaro, E., Gámez, M., and García, N. (2007). Multiclass corporate failure prediction by adaboost.m1. *Internatioal Advances in Economic Research*, 13(3): 301–312.

Alfaro, E., Gámez, M., and García, N. (2008a). Linear discriminant analysis versus adaboost for failure forecasting. *Spanish Journal of Finance and Accounting/Revista Española de Financiación y Contabilidad*, 37(137): 13–32.

Alfaro, E., García, N., Gámez, M., and Elizondo, D. (2008b). Bankruptcy forecasting: An empirical comparison of adaboost and neural networks. *Decision Support Systems*, 45: 110–122.

Alfaro, E., Gámez, M., and García, N. (2013). adabag: An R package for classification with boosting and bagging. *Journal of Statistical Software*, 54: 1–35.

Alfaro, E., Gámez, M., and García, N. (2014). *Combinación de Árboles de Clasificación con R. Aplicación a la Economía*. Alfa Delta Digital (ADD), ISBN: 978-84-9075-903-5, Valencia.

Alfaro, E., Alfaro, J. L., Gámez, M., and García, N. (2015). A comparison of different classification techniques to determine the change causes in hotelling's t2 control chart. *Quality and Reliability Engineering International*, 31(7): 1255–1263. QRE-14-0424.R3.

Alfaro, E., Gamez, M., and Garcia, N. (2018). Adabag: Applies multiclass adaboost. m1, adaboost-samme and bagging. *R Package Version*, 4.2.

Allen, R. G., Pereira, L. S., Raes, D., and Smith, M. (1998). *Crop evapotranspiration-guidelines for computing crop water requirements*. FAS irrigation and drainage paper 56. FAO, Rome, 300(9): D05109.

Altman, E. I. (1968). Financial ratios, discriminant analysis and the prediction of corporate bankruptcy. *Journal of Finance*, 23(4): 589–609.

Banfield, R. E., Hall, L. O., Bowyer, K. W., and Kegelmeyer, W. P. (2007). A comparison of decision tree ensemble creation niques. *IEEE Transactions on Pattern Analysis and Machine Intelligence*, 29(1): 173–180.

Bartlett, P. L., Jordan, M. I., and McAuliffe, J. D. (2006). Convexity, classification and risk bounds. *Journal of the American Statistical Association*, 101: 138–156.

Bauer, E. and Kohavi, R. (1999). An empirical comparison of voting classification algorithm: Bagging, boosting and variants. *Machine Learning*, 36: 105–139.

Ensemble Classification Methods with Applications in R, First Edition.
Edited by Esteban Alfaro, Matías Gámez, and Noelia García.

Bay, S. D. (1999). Nearest neighbor classification from multiple feature subsets. *Intelligent Data Analysis*, 3(3): 191–209.

Beaver, W. H. (1966). Financial ratios as predictors of failure. *Journal of Accounting Research*, 71–111.

Becu, N., Sangkapitux, C., Neef, A., Kitchaicharoen, J., and Elstner, P. (2006). Participatory simulation sessions to support collective decision: the case of water allocation between a thai and a hmong village in northern thailand. In *Proceedings of the International Symposium Towards Sustainable Livelihoods and Ecosystems in Moutainous Regions, Chiang Mai, Thailand*, pp. 7–9.

Berg, D. (2007). Bankruptcy prediction by generalized additive models. *Applied Stochastic Models in Business and Industry*, 23(2): 129–143.

Bischl, B., Lang, M., Kotthoff, L., Schiffner, J., Richter, J., Studerus, E., Casalicchio, G., and Jones, Z. M. (2016). MLR: Machine learning in R. *Journal of Machine Learning Research*, 17(170): 1–5.

Bishop, C. M. (1995). *Neural networks for pattern recognition*. Oxford University Press.

Bontemps, C. and Couture, S. (2002). Irrigation water demand for the decision maker. *Environment and Development Economics*, 7(4): 643–657.

Breiman, L. (1996a). Bagging predictors. *Machine Learning*, 24(2): 123–140.

Breiman, L. (1996b). Bias, variance, and arcing classifiers. Technical Report 460, Statistics Department, University of California at Berkeley.

Breiman, L. (1998). Arcing classifiers. *Annals of Statistics*, 26(3): 801–849.

Breiman, L. (1999). Random Forests – random features. Technical Report 567, Statistics Department, University of California at Berkeley.

Breiman, L. (2001). Random forest. *Machine Learning*, 45(1): 5–32.

Breiman, L., Friedman, J. H., Olshenn, R., and Stone, C. J. (1984). *Classification and Regression Trees*. Wadsworth International Group, Belmont.

Cardoso, J. and da Costa, J. P. (2007). Learning to classify ordinal data: The data replication method. *Journal of Machine Learning Research*, 8: 1393–1429.

Chang, C. and Lin, C. (2011). Libsvm: A library for support vector machines. *ACM transactions on intelligent systems and technology (TIST)*, 2: 27. http://www.csie.ntu.edu.tw/~cjlin/libsvm.

Charalambous, C., Charitou, A., and Kaourou, F. (2000). Comparative analysis of artificial neural network models: Application in bankruptcy prediction. *Annals of Operations Research*, 99(1–4): 403–425.

Chen, T. and Guestrin, C. (2016). Xgboost: A scalable tree boosting system. In *Proceedings of the 22nd ACM SIGKDD International Conference on Knowledge Discovery and Data Mining*, KDD '16, pp. 785–794. ACM, New York.

Chen, T., He, T., Benesty, M., Khotilovich, V., and Tang, Y. (2017). *xgboost: Extreme Gradient Boosting*. R package version 0.6-4.

Chrzanowska, M. and Witkowska, D. (2007a). Application of selected methods to the individual borrowers recognition. *Taksonomia, Classification and Data Analysis: Theory and Application*, 14: 108–114.

Chrzanowska, M. and Witkowska, D. (2007b). Classification trees in the borrowers recognition. *Quantitative Methods in Economic Research No. VIII*, pp. 291–30. Warsaw Agricultural University SGGW Press, Warszawa.

Chrzanowska, M. and Witkowska, D. (2009). Investigation of the stability of classification trees solutions obtained for the different structure of the borrowers sample.

In *Mathematical, Econometrical and Computational Methods in Finance and Insurance*, pp. 63–72. University of Economics in Katowice.

Chrzanowska, M., Alfaro, E., and Witkowska, D. (2009a). The individual borrowers recognition: Single and ensemble trees. *Expert Systems with Applications*, 36(3, Part 2): 6409–6414.

Chrzanowska, M., Alfaro, E., and Witkowska, D. (2009b). Study on the stability of ensemble trees: Example of the polish credit scoring application. *Polish Journal of Environmental Studies*, 18(5B): 63–67.

Chu, W. and Keerthi, S. (2007). Support Vector Ordinal Regression. *Neural Computation*, 19: 792–815.

Clémençon, S., Lugosi, G., and Vayatis, N. (2005). Ranking and scoring using empirical risk minimization. In *Proceedings of COLT*, pp. 1–15. Springer-Verlag, Berlin/Heidelberg.

Clémençon, S., Lugosi, G., and Vayatis, N. (2008). Ranking and empirical minimization of *U*-statistics. *Annals of Statistics*, 36: 844–874.

Clements, M. S., Armstrong, B. K., and Moolgavkar, S. H. (2005). Lung cancer rate predictions using generalized additive models. *Biostatistics*, 6(4): 576–589.

Cohen, W. W. (1995). Fast effective rule induction. In *Proceedings of the 12th International Conference on Machine Learning*, pp. 115–123. Morgan Kaufmann Publishers.

Cohen, W. W. and Singer, Y. (1999). A simple, fast, and effective rule learner. In *Proceedings of the 16th Conference of American Association for Artificial Intelligence*, pp. 335–342. American Association for Artificial Intelligence.

Condorcet, M. D. (1785). *Essai sur l'application de l'analyse á la probabilité des décisions: rendues á la pluralité des voix*. De l'imprimerie royale, París.

Cortez, P. (2016). *rminer: Data Mining Classification and Regression Methods*. R package version 1.4.2.

Cossock, D. and Zhang, T. (2006). Subset ranking using regression. In *Proceedings of COLT*, pp. 605–619. Springer-Verlag, Berlin/Heidelberg.

Coussement, K. and De Bock, K. W. (2013). Customer churn prediction in the online gambling industry: The beneficial effect of ensemble learning. *Journal of Business Research*, 66(9): 1629–1636.

Cullinan, G. J. (1977). Picking them by their batting averages' recency-frequency-monetary method of controlling circulation. Manual release, 2103.

Culp, M., Johnson, K., and Michailides, G. (2006). ada: An R package for stochastic boosting. *Journal of Statistical Software*, 17(2): 1–27.

De Bock, K. W. and Van den Poel, D. (2011). An empirical valuation of rotation-based ensemble classifiers for customer churn prediction. *Expert Systems with Applications*, 38(10): 12293–12301.

De Bock, K. W., Coussement, K., and Van den Poel, D. (2010). Ensemble classification based on generalized additive models. *Computational Statistics and Data Analysis*, 54(6): 1535–1546.

De Bock, K., Coussement, K., and Van den Poel, D. (2018). Gamens: Applies gamens. *GAMrsm and GAMbag ensemble classifiers, R Package version*, 1.2.1

de Llano Monelos, P., Piñeiro Sánchez, C., and Rodríguez López, M. (2016). Predicción del fracaso empresarial: Una contribución a la síntesis de una teoría mediante el análisis comparativo de distintas técnicas de predicción. *Estudios de Economía*, 43(2): 163–198.

Demsar, J. (2006). Statistical comparisons of classifiers over multiple data sets. *Journal of Machine Learning Research*, 7: 1–30.

Deng, W., Zheng, Q., Lian, S., Chen, L., and Wang, X. (2010). Ordinal Extreme Learning Machine. *Neurocomputing*, 74(1–3): 447–456.

Dietterich, T. G. (2000a). Ensemble methods in machine learning. In *Multiple Classifier Systems*, volume 1857 of *Lecture Notes in Computer Science*, pp. 1–15. Springer, Berlin/Heidelberg.

Dietterich, T. G. (2000b). An experimental comparison of three methods for constructing ensembles of decision trees: Bagging, boosting, and randomization. *Machine Learning*, 40(2): 139–157.

Drucker, H. and Cortes, C. (1996). Boosting decision trees. In *Advances in neural information processing systems*, pp. 479–485. MIT Press, Cambridge, MA.

Dua, D. and Karra Taniskidou, E. (2017). UCI Machine Learning Repository [http://archive.ics.uci.edu/ml]. University of California, School of Information and Computer Science, Irvine, CA.

du Jardin, P. (2017). Dynamics of firm financial evolution and bankruptcy prediction. *Expert Systems with Applications*, 75: 25–43.

Efron, B. (1982). *The jackknife, the bootstrap and other resampling plans*. SIAM.

Efron, B. (1986). How biased is the apparent error rate of a prediction rule? *Journal of the American Statistical Association*, 81(394): 461–470.

Efron, B. and Tibshirani, R. J. (1994). *An introduction to the bootstrap*. CRC Press.

Efron, B. and Tibshirani, R. J. (1995). *Cross-validation and the bootstrap: Estimating the error rate of a prediction rule*. Division of Biostatistics, Stanford University.

Eibl, G. and Pfeiffer, K. P. (2002). How to make adaboost. m1 work for weak base classifiers by changing only one line of the code. In *European Conference on Machine Learning*, pp. 72–83. Springer.

Eibl, G. and Pfeiffer, K.-P. (2005). Multiclass boosting for weak classifiers. *Journal of Machine Learning Research*, 6(Feb): 189–210.

Emerson, L., Greer-Walker, M., and Witthames, P. (1990). A stereological method for estimating fish fecundity. *Journal of Fish Biology*, 36: 721–730.

Fernández-Delgado, M., Cernadas, E., Barro, S., Ribeiro, J., and Neves, J. (2014a). Direct Kernel Perceptron (DKP): ultra-fast kernel ELM-based classification with non-iterative closed-form weight calculation. *Neural Networks*, 50: 60–71.

Fernández-Delgado, M., Cernadas, E., Barro, S., and Amorim, D. (2014b). Do we need hundreds of classifiers to solve real classification problems? *Journal of Machine Learning Research*, 15: 3133–3181.

Ferrando Bolado, M. and Blanco Ramos, F. (1998). La previsión del fracaso empresarial en la comunidad valenciana: aplicación de los modelos discriminante y logit. *Spanish Journal of Finance and Accounting/Revista Española de Financiación y Contabilidad*, 27(95): 499–540.

Fisher, R. A. (1936). The use of multiple measurements in taxonomic problems. *Annals of Human Genetics*, 7(2): 179–188.

Frank, F. and Hall, M. (2001). A simple approach to ordinal classification. Proceedings of the 12th European Conference on Machine Learning. *Lecture Notes in Computer Science*, 2167: 145–156.

Freund, Y. (1995). Boosting a weak learning algorithm by majority. *Information and Computation*, 121(2): 256–285.

Freund, Y. and Schapire, R. E. (1996). Experiments with a new boosting algorithm. In *Proceedings of the Thirteenth International Conference on Machine Learning*, pp. 148–156. Morgan Kaufmann, Bari (Italy).

Freund, Y. and Schapire, R. E. (1997). A decision-theoretic generalization of on-line learning and an application to boosting. *Journal of Computer and System Sciences*, 55(1): 119–139.

Freund, Y. and Schapire, R. E. (1998). Discussion: Arcing classifiers. *Annals of Statistics*, 26(3): 824–832.

Freund, Y., Iyer, R., Schapire, R. E., and Singer, Y. (2003). An efficient boosting algorithm for combining preferences. *Journal of Machine Learning Research*, 4(Nov): 933–969.

Friedman, M. (1937). The use of ranks to avoid the assumption of normality implicit in the analysis of variance. *Journal of the American Statistical Association*, 32(200): 675–701.

Friedman, M. (1940). A comparison of alternative tests of significance for the problem of m rankings. *Annals of Mathematical Statistics*, 11(1): 86–92.

Friedman, J. H. and Popescu, B. E. (2008). Predictive learning via rule ensembles. *Annals of Applied Statistics*, 916–954.

Friedman, J. H., Hastie, T., and Tibshirani, R. (2000). Additive logistic regression: A statistical view of boosting. *Annals of Statistics*, 38(2): 337–407.

Frydman, H., Altman, E. I., and Kao, D.-L. (1985). Introducing recursive partitioning for financial classification: The case of financial distress. *Journal of Finance*, 40(1): 269–91.

Fumera, G. and Roli, F. (2001). Error rejection in linearly combined multiple classifiers. In *International Workshop on Multiple Classifier Systems*, pp. 329–338. Springer.

Fürnkranz, J. (1999). Separate-and-conquer rule learning. *Artificial Intelligence Review*, 13(1): 3–54.

Gabás, F. (1990). Técnicas actuales de análisis contable. evaluación de la solvencia empresarial. *Instituto de Contabilidad y Auditoría de Cuentas (ICAC)*.

Gallego, A. and Gómez, J. (2002). Efectos marginales en la metodología multilogit: interpretación y contrastación empírica en los modelos de insolvencia. In Doldán, F. and Rodríguez, M. (editors), *La gestión del riesgo de crédito. Métodos y Modelos de Predicción de la Insolvencia Empresarial*, pp. 253–288. AECA, Madrid.

González-Rufino, E., Carrión, P., Cernadas, E., Fernández-Delgado, M., and Domínguez-Petit, R. (2013). Exhaustive comparison of colour texture features and classification methods to discriminate cells categories in histological images of fish ovary. *Pattern Recognition*, 46: 2391–2407.

Górecki, T. and Krzyśko, M. (2015). Regression methods for combining multiple classifiers. *Communications in Statistics – Simulation and Computation*, 44(3): 739–755.

Greenwell, B. M. (2017). pdp: An r package for constructing partial dependence plots. *The R Journal*, 9(1): 421–436.

Guo, L. and Boukir, S. (2013). Margin-based ordered aggregation for ensemble pruning. *Pattern Recognition Letters*, 34(6): 603–609.

Guyon, I., Gunn, S., Nikravesh, M., and Zadeh, L. (2006). *Feature Extraction: Foundations and Applications. Pattern Recognition*. Springer.

Hansen, L. K. and Salamon, P. (1990). Neural network ensembles. *IEEE Transactions on Pattern Analysis and Machine Intelligence*, 12(10): 993–1001.

Hastie, T. and Tibshirani, R. (1986). Generalized additive models. *Statistical Science*, 1(3): 297–310.

Hastie, T. and Tibshirani, R. (1990). *Generalized additive models*. Wiley Online Library.

Haykin, S. (1994). *Neural networks: a comprehensive foundation*. Prentice Hall PTR.

Hill, T. and Lewicki, P. (2007). *Statistics: Methods and Applications*. StatSoft, Tulsa.

Ho, T. K. (1998). The random subspace method for constructing decision forests. *IEEE Transactions on Pattern Analysis and Machine Intelligence*, 20(8): 832–844.

Hoeting, J. A., Madigan, D., Raftery, A. E., and Volinsky, C. T. (1999). Bayesian model averaging: a tutorial. *Statistical Science*, 14: 382–401.

Holm, S. (1979). A simple sequentially rejective multiple test procedure. *Scandinavian Journal of Statistics*, 6(2): 65–70.

Hothorn, T., Buhlmann, P., Kneib, T., Schmid, M., and Hofner, B. (2017). mboost: *Model-Based Boosting*. R package version 2.8.

Huang, G., Wang, D., and Lan, Y. (2011). Extreme Learning Machines: a survey. *International Journal of Machine Learning and Cybernetics*, 2: 107–122.

Kastellec, J. P. (2010). The statistical analysis of judicial decisions and legal rules with classification trees. *Journal of Empirical Legal Studies*, 7(2): 202–230.

Kawakita, M., Minami, M., Eguchi, S., and Lennert-Cody, C. (2005). An introduction to the predictive technique adaboost with a comparison to generalized additive models. *Fisheries Research*, 76(3): 328–343.

Kim, H.-S. and Yoon, C.-H. (2004). Determinants of subscriber churn and customer loyalty in the korean mobile telephony market. *Telecommunications Policy*, 28(9): 751–765.

Kittler, J., Hatef, M., Duin, R. P., and Matas, J. (1998). On combining classifiers. *IEEE Transactions on Pattern Analysis and Machine Intelligence*, 20(3): 226–239.

Kohavi, R. (1995). A study of cross-validation and bootstrap for accuracy estimation and model selection. In *Proceedings of the International Joint Conference on Artificial Intelligence*, vol. 14, pp. 1137–1145. Lawrence Erlbaum Associates Ltd.

Kohavi, R. and Kunz, C. (1997). Option decision trees with majority votes. In *14th International Conference on Machine Learning*, vol. 97, pp. 161–169. Morgan Kaufmann Publishers.

Krempl, G. and Hofer, V. (2008). Partitioner trees: Combining boosting and arbitrating. In Okun, O. and Valentini, G., editors, *2nd Workshop SUEMA 2008 (ECAI 2008)*, pp. 61–66. Springer.

Krogh, A. and Vedelsby, J. (1995). Neural network ensembles, cross validation, and active learning. In Touretzky, D.S., Mozer, M.C. and Hasselmo, M.E., editors, *Advances in neural information processing systems*, pp. 231–238. Morgan Kaufmann Publishers.

Kubat, M. and Matwin, S. (1997). Addressing the curse of imbalanced training sets: one-sided selection. In *International Conference on Machine Learning*, vol. 97, pp. 179–186. Morgan Kaufmann Publishers, Nashville.

Kubus, M. (2008). Zastosowanie metody boosting w indukcji reguł. *Taksonomia. Klasyfikacja i analiza danych-teoria i zastosowania*, 15(7(1207)): 470–477.

Kuhn, M. (2008). Building predictive models in R using the caret package. *Journal of Statistical Software*, 28(5): 1–26.

Kuhn, M. (2017). *caret: Classification and Regression Training*. R package version 6.0-77. Contributions from Jed Wing, Steve Weston, Andre Williams, Chris Keefer, Allan Engelhardt and others.

Kumar, P. R. and Ravi, V. (2007). Bankruptcy prediction in banks and firms via statistical and intelligent techniques–a review. *European Journal of Operational Research*, 180(1): 1–28.

Kuncheva, L. I. (2004). *Combining Pattern Classifiers: Methods and Algorithms.* Wiley-Interscience.

Lado, M. J., Cadarso-Suarez, C., Roca-Pardinas, J., and Tahoces, P. G. (2006). Using generalized additive models for construction of nonlinear classifiers in computer-aided diagnosis systems. *IEEE Transactions on Information Technology in Biomedicine*, 10(2): 246–253.

Lahsasna, A., Ainon, R. N., and Teh, Y. W. (2010). Credit scoring models using soft computing methods: A survey. *International Arab Journal of Information Technology*, 7(2): 115–123.

Lam, L. (2000). Classifier combinations: implementations and theoretical issues. In *International Workshop on Multiple Classifier Systems*, pp. 77–86. Springer.

Langley, P. (2000). Crafting papers on machine learning. In *International Conference on Machine Learning*, pp. 1207–1216. Morgan Kaufmann Publishers.

Le Bars, M., Attonaty, J.-M., Pinson, S., and Ferrand, N. (2005). An agent-based simulation testing the impact of water allocation on farmers' collective behaviors. *Simulation*, 81(3): 223–235.

Lemmens, A. and Croux, C. (2006). Bagging and boosting classification trees to predict churn. *Journal of Marketing Research*, 43(2): 276–286.

Li, L. (2006). Multiclass boosting with repartitioning. In *Proceedings of the 23rd International Conference on Machine Learning*, pp. 569–576. ACM.

Li, L. and Lin, H. (2007). Ordinal regression by extended binary classification. In *Advances in Neural Information Processes and Systems 19*, pp. 865–872.

Liaw, A. and Wiener, M. (2002). Classification and regression by randomforest. *R News*, 2(3): 18–22.

Lin, H. and Li, L. (2006). Large-margin thresholded ensembles for ordinal regression: Theory and practice. In *Algorithmic Learning Theory*, volume 4264 of *Lecture Notes in Computer Science*, pp. 319–333. Springer, Berlin/Heidelberg.

Lippmann, R. P. (1991). A critical overview of neural network pattern classifiers. In *Proceedings of the IEEE Workshop on Neural Networks for Signal Processing*, pp. 266–275. IEEE.

Louzada, F., Anacleto-Junior, O., Candolo, C., and Mazucheli, J. (2011). Poly-bagging predictors for classification modelling for credit scoring. *Expert Systems with Applications*, 38(10): 12717–12720.

Maindonald, J. and Braun, J. (2010). *Data Analysis and Graphics Using R*, 3rd edition. Cambridge University Press.

McCullagh, P. and Nelder, J. A. (1989). *Generalized Linear Models*. Chapman & Hall/CRC.

Mendelson, S. (2002). Improving the sample complexity using global data. *IEEE Transaxtions on Information Theory*, 48: 1977–1991.

Michalski, R. S. (1969). On the quasi-minimal solution of the general covering problem. In *Proceedings of the 5th International Symposium on Information Processing (FCIP-69)*, volume A3 (Switching Circuits), pp. 125–128. Bled (Yugoslavia).

Momparler, A., Carmona, P., and Climent, F. (2016). Banking failure prediction: a boosting classification tree approach. *Spanish Journal of Finance and Accounting/Revista Española de Financiación y Contabilidad*, 45(1): 63–91.

Mukherjee, I. and Schapire, R. E. (2010). Learning with continuous experts using drifting games. *Theoretical Computer Science*, 411(29): 2670–2683. Algorithmic Learning Theory (ALT 2008).

Mukherjee, I. and Schapire, R. E. (2011). A theory of multiclass boosting. *Advances in Neural Information Processing Systems*, 23(4): 1722–1722.

Ohlson, J. A. (1980). Financial ratios and the probabilistic prediction of bankruptcy. *Journal of Accounting Research*, 18(1): 109–131.

Ojala, T., Piatikäinen, M., and Mäenpää, T. (2002). Multiresolution grey-scale and rotation invariant texture classification with local binary pattern. *IEEE Transactions on Pattern Analysis and Machine Intelligence*, 24 (7): 971–987.

Pérez, M. (2006). Artificial neural networks and bankruptcy forecasting: a state of the art. *Neural Computing & Applications*, 15(2): 154–163.

Pérez-Ortiz, M., Gutiérrez, P., and Hervás-Martínez, C. (2014). Projection based ensemble learning for ordinal regression. *IEEE Transactions on Cybernetics*, 44(5): 681–694.

Pérez-Ortiz, M., Fernández-Delgado, M., Cernadas, E., Domínguez-Petit, R., Gutiérrez, P., and Hervás, C. (2015). On the use of nominal and ordinal classifiers for the discrimination of states of development in fish oocytes. *Neural Processing Letters*, 46(9): 1–16.

Pfahringer, B., Holmes, G., and Wang, C. (2004). Millions of random rules. In *Proceedings of the Workshop on Advances in Inductive Rule Learning, 15th European Conference on Machine Learning (ECML)*. Knowledge Engineering Group, Pisa.

Pino-Mejias, R., Carrasco-Mairena, M., Pascual-Acosta, A., Cubiles-De-La-Vega, M.-D., and Munoz-Garcia, J. (2008). A comparison of classification models to identify the fragile x syndrome. *Journal of Applied Statistics*, 35(3): 233–244.

Polikar, R. (2006). Ensemble based systems in decision making. *IEEE Circuits and Systems Magazine*, 6(3): 21–45.

Pollard, D. (1984). *Convergence of Stochastic Processes*. Springer, New York.

Provost, F. J., Fawcett, T., Kohavi, R. (1998). The case against accuracy estimation for comparing induction algorithms. In *Fifteenth International Conference on Machine Learning*, vol. 98, pp. 445–453. Morgan Kaufmann Publishers.

Quinlan, J. R. (1986). Induction of decision trees. *Machine Learning*, 1(1): 81–106.

Quinlan, J. R. (1993). *C4. 5: Programs for machine learning*. Morgan Kaufmann, San Mateo.

Quinlan, J. R. (1996). Boosting first-order learning. In *International Workshop on Algorithmic Learning Theory*, pp. 143–155. Springer.

Quinlan, J.R. and Rivest, L. (1989). Inferring decision trees using the minimum description length principle. *Information and Computation*, 80(3): 227–248.

R Development Core Team (2017). *R: A Language and Environment for Statistical Computing*. R Foundation for Statistical Computing, Vienna. ISBN 3-900051-07-0.

Rejchel, W. (2012). On ranking and generalization bounds. *Journal of Machine Learning Research*, 13: 1373–1392.

Rejchel, W. (2015). Fast rates for ranking with large families. *Neurocomputing*, 168: 1104–1110.

Ridgeway, G. (2017). *GBM: Generalized Boosted Regression Models*. R package version 2.1-3.

Rissanen, J. (1978). Modeling by shortest data description. *Automatica*, 14(5): 465–471.

Rodriguez, M. (2009). Multi-class boosting. Technical report, Notes on AdaBoost algorithms, Department of Computer Science, University of California, Santa Cruz.

Rodriguez, J. J., Kuncheva, L. I., and Alonso, C. J. (2006). Rotation forest: A new classifier ensemble method. *IEEE Transactions on Pattern Analysis and Machine Intelligence*, 28(10): 1619–1630.

Rodríguez López, M. (2001). Predicción del fracaso empresarial en compañías no financieras. consideración de técnicas de análisis multivariante de corte paramétrico. *Actualidad Financiera*, 6(6): 27–42.

Rudin, C. (2006). Ranking with a p-norm push. In *Proceedings of COLT*, pp. 589–604. Springer-Verlag, Berlin/Heidelberg.

Schapire, R. E. (1990). The strength of weak learnability. *Machine Learning*, 5(2): 197–227.

Schapire, R. E. (1999). A brief introduction to boosting. In *International Joint Conference on Artificial Intelligence*, vol. 16, pp. 1401–1406. IJCAI Organization.

Schapire, R. E. and Freund, Y. (2012). *Boosting: Foundations and Algorithms*. MIT Press, Cambridge, MA.

Schapire, R. E. and Singer, Y. (1999). Improved boosting algorithms using confidence-rated predictions. *Machine Learning*, 37(3): 297–336.

Schapire, R. E., Freund, Y., Bartlett, P., and Lee, W. S. (1998). Boosting the margin: A new explanation for the effectiveness of voting methods. *Annals of Statistics*, 26: 322–330.

Serrano-Cinca, C. (1997). Feedforward neural networks in the classification of financial information. *European Journal of Finance*, 3(3): 183–202.

Skurichina, M. (2001). *Stabilizing weak classifiers: Regularization and combining techniques in discriminant analysis*. Ph.D. thesis, Delft University of Technology.

Skurichina, M. and Duin, R. (2001). *Bagging and the random subspace method for redundant feature spaces*, pp. 1–10. Springer.

Specht, D. (1990). Probabilistic Neural Networks. *Neural Networks*, 3: 109–118.

Stewart, B. M. and Zhukov, Y. M. (2009). Use of force and civil-military relations in Russia: An automated content analysis. *Small Wars and Insurgencies*, 20(2): 319–343.

Strobl, C., Boulesteix, A.-L., Zeileis, A., and Hothorn, T. (2007). Bias in random forest variable importance measures: Illustrations, sources and a solution. *BMC Bioinformatics*, 8(1): 25.

Sun, Y., Kamel, M. S., Wong, A. K., and Wang, Y. (2007). Cost-sensitive boosting for classification of imbalanced data. *Pattern Recognition*, 40(12): 3358–3378.

Sun, B., J. L., Wu, D., Zhang, S., and Li, W. (2010). *IEEE Transactions on Knowledge and Data Engineering*, 22: 906–910.

Swets, J. A. (1988). Measuring the accuracy of diagnostic systems. *Science*, 240(4857): 1285–1293.

Therneau, T. M. and Atkinson, B. (2017). Rpart: *Recursive Partitioning*. R package version 4.1-11.

Tibshirani, R. (1996a). Bias, variance and prediction error for classification rules. Technical report, Department of Statistics, University of Toronto.

Tibshirani, R. (1996b). Regression shrinkage and selection via the lasso. *Journal of the Royal Statistical Society, Series B*, 58: 267–288.

Tkáč, M. and Verner, R. (2016). Artificial neural networks in business: Two decades of research. *Applied Soft Computing*, 38: 788–804.

Torgo, L. (2010). *Data Mining with R: Learning with Case Studies*. Data Mining and Knowledge Discovery Series. Chapman & Hall/CRC.

Valentini, G. and Masulli, F. (2002). Ensembles of learning machines. In *Italian Workshop on Neural Nets*, pp. 3–20. Springer.

van der Spoel, S. (2016). *crtests: Classification and Regression Tests.* R package version 0.2.1.

van der Vaart, A. W. and Wellner, J. A. (1996). *Weak Convergence and Empirical Processes: With Applications to Statistics.* Springer Verlag, New York.

Vapnik, V. N. (1998). *Statistical Learning Theory.* Wiley, New York.

Wang, G., Ma, J., Huang, L., and Xu, K. (2012). Two credit scoring models based on dual strategy ensemble trees. *Knowledge-Based Systems,* 26: 61–68.

Warmuth, M. K., Glocer, K. A., and Rätsch, G. (2008). Boosting algorithms for maximizing the soft margin. In *Advances in neural information processing systems,* vol. 20, pp. 1585–1592.

Weiss, S. M. and Kulikowski, C. A. (1991). *Computer systems that learn: classification and prediction methods from statistics, neural nets, machine learning, and expert systems.* Morgan Kaufmann Publishers Inc.

White, G., Sondhi, A., and Fried, D. (2003). *The analysis and use of financial statements.* John Wiley & Sons.

Wilcoxon, F. (1946). Individual comparisons of grouped data by ranking methods. *Journal of Economic Entomology,* 39(2): 269–270.

Wilson, R. L. and Sharda, R. (1994). Bankruptcy prediction using neural networks. *Decision Support Systems,* 11(5): 545–557.

Wolpert, D. H. and Macready, W. G. (1999). An efficient method to estimate bagging's generalization error. *Machine Learning,* 35(1): 41–55.

Wright, J. L. (1982). New evapotranspiration crop coefficients. *Journal of the Irrigation and Drainage Division,* 108(IR2): 57–74.

Wu, S. and Nagahashi, H. (2015). Analysis of generalization ability for different adaboost variants based on classification and regression trees. *Journal of Electrical and Computer Engineering,* 2015: 1–17.

Zhang, C.-X. and Zhang, J.-S. (2008). Rotboost: A technique for combining rotation forest and adaboost. *Pattern Recognition Letters,* 29(10): 1524–1536.

Zhou, L. (2013). Performance of corporate bankruptcy prediction models on imbalanced dataset: The effect of sampling methods. *Knowledge-Based Systems,* 41: 16–25.

Zhou, L. and Lai, K. K. (2017). Adaboost models for corporate bankruptcy prediction with missing data. *Computational Economics,* 50(1): 69–94.

Zhu, J., Zou, H., Rosset, S., and Hastie, T. (2009). Multi-class adaboost. *Statistics and Its Interface,* 2: 349–360.

Zięba, M., Tomczak, S. K., and Tomczak, J. M. (2016). Ensemble boosted trees with synthetic features generation in application to bankruptcy prediction. *Expert Systems with Applications,* 58: 93–101.

Zmijewski, M. E. (1984). Methodological issues related to the estimation of financial distress prediction models. *Journal of Accounting Research,* 22: 59–82.

Index

Ensemble Classification Methods with Applications in R, First Edition.
Edited by Esteban Alfaro, Matías Gámez, and Noelia García.